BOWLING GREEN STATE UNIVERSITY
DISCARDED
LIBRARY

COMPUTATIONAL QUANTUM CHEMISTRY

COMPUTATIONAL QUANTUM CHEMISTRY

ALAN HINCHLIFFE BSc, PhD, DSc
Chemistry Department
UMIST
(University of Manchester Institute of Science and Technology)

JOHN WILEY & SONS

Chichester • New York • Brisbane • Toronto • Singapore

Library of Congress Cataloging in Publication Data:

Hinchliffe, Alan.
Computational quantum chemistry/by Alan Hinchliffe.
p. cm.
ISBN 0 471 91716 8
1. Quantum chemistry—Data processing. I. Title.
QD462.6.E43H56 1989
541.2′8′0285—dc 19 88–14405
CIP

British Library Cataloging in Publication Data:

Hinchliffe, Alan
Computational quantum chemistry
1. Chemistry. Quantum theory
I. Title
541.2′8

ISBN 0 471 91716 8

Typeset by Best-set Typesetter Ltd.,
Printed and bound in Great Britain by Anchor Brendon Ltd. Tiptree, Essex

Contents

Preface

As a chemistry undergraduate in the 1960s, I learned quantum chemistry as a very 'theoretical' subject. In order to get to grips with the colour of carrots I knew that I had somehow to understand

$$|\int \Psi_k{}^* \Sigma \mathbf{r}_i \Psi_0 \mathrm{d}\tau|^2$$

but I didn't really know how to actually calculate the quantity, or have the slightest idea as to what the answer would be. Even the simplest calculations such as $\int x^2 \exp(-x^2)\,\mathrm{d}x$ or a Huckel MO calculation on butadiene seemed to be the limits of knowledge. The basic chemical concepts disappeared in a mass of algebra, and I suspect that most students who have struggled to master the simplest possible problem in quantum chemistry (a hydrogen atom) will agree.

In those days, numerical analysis was done with pencils and large sheets of paper, and computers were only available in Lyons Tea Shops (true!)

Today we live in a world where everything from the chairs we sit in to the cars we drive are firstly designed by computer simulation and then built. There is no reason why chemistry should not be part of such a world, and why it should not be seen to be part of such a world by chemistry undergraduates.

The electronic structure and physical properties of any molecule in any of its stationary states can be determined in principle by solution of the time-independent Schrödinger equation, and this apparently simple fact has been appreciated since the birth of modern quantum theory in 1926. Only in the last 20 years, with the advent of computers, has it become possible to solve the necessary equations without recourse to severe approximations and dubious 'simplifications'.

The label 'Computational Quantum Chemistry' has been coined to describe the application of quantum-mechanical techniques to problems involving molecular structure and properties. The emergence over the last two decades of computational quantum chemistry as a major tool in industry has been great enough to warrant an article in the 23 August 1983 edition of the *Wall Street Journal*.

Computational quantum chemistry is now an established part of the chemist's armoury. It can be used as an analytical tool in the same sense that an nmr spectrometer or X-ray diffractometer can be used to rationalize the structure of a known molecule. Its true place, however, is as a predictive tool, to be considered before the experiment. It should be considered when asking the question 'What if...?'.

I was asked a small number of years ago to write a book aimed at research workers who wished to use the techniques of computational quantum chemistry in order to solve problems. I envisaged that consumers of the book, *Ab Initio Determination of Molecular Properties*, would have an understanding of the concepts of quantum chemistry and access to a suitable software package. Part of my brief was to emphasize the likely accuracy of the calculations.

At undergraduate level the scene is also changing fast. Our present generation of chemistry students are much more computer literate than the previous generation. At UMIST we teach 'computer appreciation' to all our chemistry undergraduates; this involves them in logging in to databases for ^{13}C nmr, for crystal structures and for *Chem. Abs.* We also expose our students to GINO, to NAG and to SPSS. Our students are by no means computer experts, but they do have an appreciation of what computers can accomplish.

We have an extremely flexible undergraduate chemistry course, and our students can specialize in their final year in a variety of well thought-out options, or they can design their own course. In particular, we run a chemical physics option in the final year, and part of the course has been 'quantum chemistry' since the chemical physics option began 19 years ago.

This book is based on the 'Computational Quantum Chemistry' module offered to our final-year chemists. Consumers are generally chemical physicists, but a growing number of 'ordinary' chemists elect for the module, presumably because of the strong computer-appreciation elements of years 1 and 2. The course is 'hands-on', and the students do not have to worry about the detailed algebra. I hope that this text will be useful for undergraduate chemistry teaching in general. A prerequisite is that the students should have attended an intermediate quantum chemistry course, and that a package such as GAUSSIAN80 should be available at the local computer centre.

In Chapter 1 I have given a summary, which for many students will be revision, of the elementary concepts needed for the remainder of the text.

Alan Hinchliffe
March 1988

CHAPTER 1

Introduction

I have assumed a knowledge of the elementary concepts of valence theory. However, it is worthwhile briefly reviewing several key concepts, and developing a systematic nomenclature and notation, so the remainder of the chapter is devoted to this aim. Most of the chapter should therefore be seen as revision.

1.1 THE SCHRÖDINGER EQUATION

Suppose we have a single particle (e.g. an electron) of mass m moving in a region of space under the influence of a potential V (for example, this potential might be provided by the nuclei in a molecule). Let us denote the position of the particle at time t by $\mathbf{r} = x\mathbf{i} + y\mathbf{j} + z\mathbf{k}$. The particle is described by a wavefunction $\Psi(\mathbf{r}, t)$ which satisfies Schrödinger's time-dependent equation

$$\left\{-\frac{\hbar^2}{2m}\left(\frac{\partial^2}{\partial x^2} + \frac{\partial^2}{\partial y^2} + \frac{\partial^2}{\partial z^2}\right) + V\right\}\Psi(\mathbf{r}, t) = i\hbar\frac{\partial\Psi(\mathbf{r}, t)}{\partial t} \tag{1.1}$$

(where $i^2 = -1$).

This partial differential equation has to be solved for $\Psi(\mathbf{r}, t)$ subject to the appropriate boundary conditions. It is usual to rewrite this equation in terms of the *gradient operator* $\boldsymbol{\nabla}$:

$$\boldsymbol{\nabla} \equiv \frac{\partial}{\partial x}\mathbf{i} + \frac{\partial}{\partial y}\mathbf{j} + \frac{\partial}{\partial z}\mathbf{k}$$

so that

$$\boldsymbol{\nabla} \cdot \boldsymbol{\nabla} = \frac{\partial^2}{\partial x^2} + \frac{\partial^2}{\partial y^2} + \frac{\partial^2}{\partial z^2}$$

and (1.1) becomes

$$\left\{-\frac{\hbar^2}{2m}\nabla^2 + V\right\}\Psi(\mathbf{r}, t) = i\hbar\frac{\partial\Psi(\mathbf{r}, t)}{\partial t} \tag{1.2}$$

In the case where the potential is independent of time, we can use a standard mathematical technique called 'separation of variables' to separate (1.1) into a *spatial* part and a *time* part: if we write $\Psi(\mathbf{r}, t) = \psi(\mathbf{r})T(t)$ then it is easy to show that

$$\left\{-\frac{\hbar^2}{2m}\nabla^2 + V\right\}\psi(\mathbf{r}) = E\psi(\mathbf{r}) \tag{1.3}$$

$$T(t) = \exp(-iEt/\hbar) \tag{1.4}$$

Equation (1.3) is usually referred to as the time-independent Schrödinger equation. We are generally interested in the time dependence if we investigate problems such as spectroscopic transitions, the response of a molecule to a rapidly oscillating electric field or scattering. We are mostly going to be concerned with the time-independent case in this book.

> *Problem 1.1*
> Substitute $\Psi(\mathbf{r}, t) = \psi(\mathbf{r})T(t)$ into expression (1.1) and rearrange to give
>
> $$\frac{1}{\psi}\left\{-\frac{\hbar^2}{2m}\nabla^2 + V\right\}\psi = \frac{i\hbar}{T}\frac{\mathrm{d}T}{\mathrm{d}t}$$

Each side of the equality involves independent variables and so each side must be equal to a constant (call it E). Hence you will find equation (1.3) and (1.4).

> *Problem 1.2*
> Solve the 'electron in a two-dimensional box' problem. The potential is zero for $x \leqslant a$ and $y \leqslant b$ but infinite everywhere else. You will find solutions as follows:
>
> $$\psi_{n,k}(x, y) = \sqrt{\frac{4}{ab}}\sin\frac{n\pi x}{a}\sin\frac{k\pi y}{b}$$
>
> $$E_{n,k} = \left(\frac{n^2}{a^2} + \frac{k^2}{b^2}\right)\frac{h^2}{8m}$$

Wavefunctions in general are complex quantities (in the mathematical sense). We will write Ψ^* as the complex conjugate of Ψ, so that the size of Ψ is given by $(\Psi^*\Psi)^{1/2}$. We will generally denote this by the symbol $|\Psi|$. Born suggested in 1926 that $|\Psi|^2$ should be best interpreted as a probability density for the particle, so that

$|\Psi|^2\,d\tau$ is the probability of finding the particle in an element of space $d\tau$ (typically $d\tau = dx\,dy\,dz$). In accordance with this interpretation, we require the wavefunction to be *normalized*, $\int |\Psi|^2\,d\tau = 1$, because there has to be a finite chance of finding the particle somewhere in space (i.e. all the probabilities have to add to 1).

1.2 THE HAMILTONIAN OPERATOR

It is usual to write equation (1.3) as

$$\hat{H}\psi(\mathbf{r}) = E\psi(\mathbf{r}) \tag{1.5}$$

where $\hat{H}$ is the *Hamiltonian operator* given by

$$\hat{H} = -\frac{\hbar^2}{2m}\nabla^2 + V \tag{1.6}$$

The time-dependent equation is then

$$\hat{H}\Psi(\mathbf{r}, t) = i\hbar\frac{\partial\Psi(\mathbf{r}, t)}{\partial t} \tag{1.7}$$

Equations such as (1.5) above, where a linear operator $\hat{H}$ acts on a function to give the function back again, play an important role in many areas of science and engineering and we call such equations *eigenvalue* equations. In general there will be a certain number of eigenvalues E_k and eigenfunctions ψ_k for which

$$\hat{H}\psi_k = E_k\psi_k$$

Sometimes there will be a finite number of eigenfunctions, sometimes an infinite number and in the latter case the eigenvalues can be discrete or continuous. We generally have to solve the differential equation for ψ and E subject to the appropriate boundary conditions.

Schrödinger's recipe allows us to construct operators. We write down the observable of interest (energy, angular momentum, electric dipole moment, etc.), in terms of position $\mathbf{r}$ and linear momentum coordinates and then substitute

$$\mathbf{p} \rightarrow -i\hbar\boldsymbol{\nabla}$$

to construct the operator.

1.3 ONE-ELECTRON ATOMS

For a one-electron atom with atomic number Z (e.g. for He^+, $Z = 2$), the Coulomb potential is given by $-Ze^2/4\pi\varepsilon_0 r$, r being the scalar distance between the electron and the nucleus. We concern ourselves with the relative motion of the electron and the nucleus, and solution of the time-independent eigenvalue problem

$$\left\{-\frac{\hbar^2}{2m}\nabla^2 - \frac{Ze^2}{4\pi\varepsilon_0 r}\right\}\psi(\mathbf{r}) = E\psi(\mathbf{r}) \tag{1.8}$$

is given in all the standard quantum chemistry texts. Because of the spherical symmetry of the potential, one rewrites ∇^2 in spherical polar coordinates and seeks a solution of the form

$$\psi(r, \theta, \phi) = R(r)\Theta(\theta)\Phi(\phi) \tag{1.9}$$

subject to the boundary conditions ($R(r) \to 0$ as $r \to \infty$, etc.). The result is the familiar set of wavefunctions, referred to as *atomic orbitals*, which are characterized by three quantum numbers n, l, m. Any mathematical function describing a single electron is referred to as an orbital.

1.4 ELECTRON SPIN

It turns out that electrons (like many other particles) have an internal angular momentum which we call 'spin'. A consequence of this property is that electrons are magnetic dipoles, and hence we have the field of magnetic resonance spectroscopy. The term *electron spin* is singularly unfortunate; electrons do *not* 'spin' on their 'axes' as they travel 'in Bohr orbits around the nucleus'. It is, however, very hard to remove this simple-minded picture from descriptive chemistry!

To describe electron spin mathematically, we introduce a spin variable s in addition to the position vector $\mathbf{r}$ and we will often write $\mathbf{x} = \mathbf{r}s$ as a composite variable. We deal with the electron spin in more detail in Chapter 2.

1.5 THE BORN–OPPENHEIMER APPROXIMATION

The simplest possible molecule is H_2^+, and we are interested in solving the time-independent Schrödinger eigenvalue equation

$$\hat{H}\Psi(\mathbf{R}_1, \mathbf{R}_2, \mathbf{r}) = E\Psi(\mathbf{R}_1, \mathbf{R}_2, \mathbf{r}) \tag{1.10}$$

where we have used $\mathbf{R}_i$ for each *nuclear* position vector and $\mathbf{r}$ for the electron position vector, as shown in Figure 1.1.

Born and Oppenheimer (1927) showed that, to a very good approximation, the motions of the nuclei and the electrons could be considered separately. There *are* phenomena such as the Renner

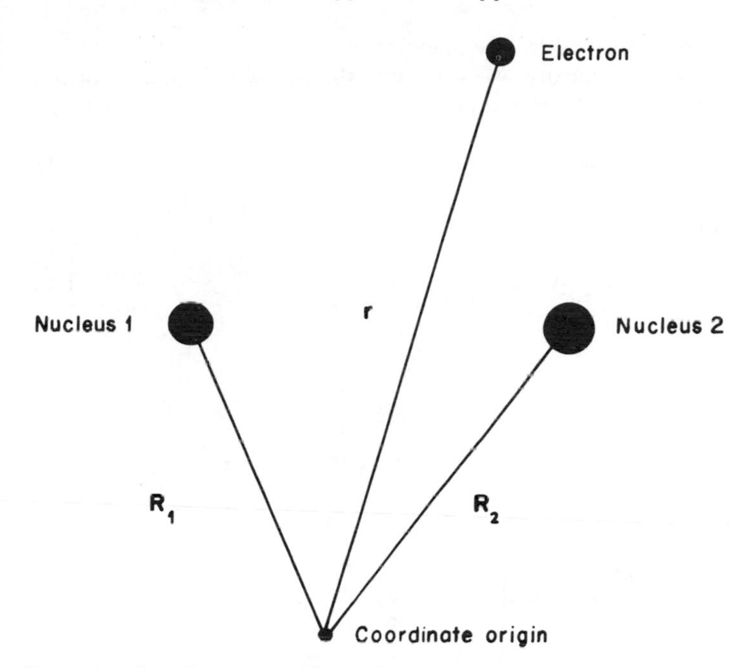

Figure 1.1 Coordinate system for treatment of H_2^+

effect where this is not true, but for 95 per cent of chemical phenomena the Born–Oppenheimer approximation is a useful one. The *electronic* wavefunction is given by

$$\hat{H}_{el}\Psi_{el}(\mathbf{r}) = E_{el}\Psi_{el}(\mathbf{r}) \tag{1.11}$$

where

$$\hat{H}_{el} = -\frac{\hbar^2}{2m}\nabla^2 - \frac{e^2}{4\pi\varepsilon_0|\mathbf{R}_1 - \mathbf{r}|} - \frac{e^2}{4\pi\varepsilon_0|\mathbf{R}_2 - \mathbf{r}|} \tag{1.12}$$

The *total* energy is

$$E_{tot} = E_{el} + \frac{e^2}{4\pi\varepsilon_0|\mathbf{R}_1 - \mathbf{R}_2|} \tag{1.13}$$

where the nuclei are clamped at $\mathbf{R}_1$, $\mathbf{R}_2$ for the purpose of calculating the electronic wavefunction.

For a polyatomic molecule consisting of n electrons and N nuclei, we are therefore interested in solving the eigenvalue equation

$$\hat{H}_{el}\Psi_{el}(\mathbf{r}_1 \ldots \mathbf{r}_n) = E_{el}\Psi_{el}(\mathbf{r}_1 \ldots \mathbf{r}_n) \tag{1.14}$$

at some *fixed* nuclear geometry.

For convenience, we will adopt the following notation for contributions to the Hamiltonian:

$$\hat{H}_{\text{el}} = \sum_{i=1}^{n} \hat{h}(i) + \frac{1}{2}\sum\sum_{i \neq j} \hat{g}(i, j) \tag{1.15}$$

where

$$\hat{h}(i) = -\frac{\hbar^2}{2m}\nabla_i^2 - \sum_{\alpha=1}^{N} \frac{Z_\alpha e^2}{4\pi\varepsilon_0|\mathbf{R}_\alpha - \mathbf{r}_i|} \tag{1.16}$$

and

$$\hat{g}(i, j) = \frac{e^2}{4\pi\varepsilon_0|\mathbf{r}_i - \mathbf{r}_j|} \tag{1.17}$$

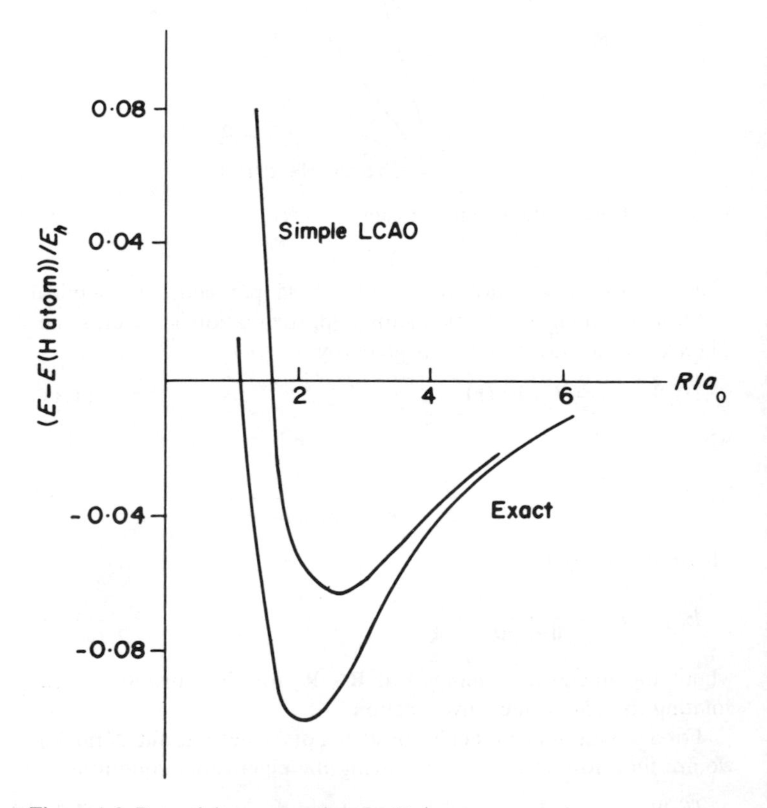

Figure 1.2 Potential energy surface for H_2^+

We often refer to the $\hat{h}(i)$ as 'one-electron operators', and in this case they comprise contributions from kinetic energy and nuclear attraction. The operator $\hat{g}(i, j)$ represents the Coulomb repulsion between electrons i and j, and we call it a 'two-electron operator'.

If we repeat the calculation of total energy for a variety of fixed nuclear positions, we will eventually calculate a *potential energy surface*. The nuclei move in a potential provided by the electron density: hence the name. For a simple diatomic molecule, such as the hydrogen molecule ion H_2^+ (to be discussed in the next section) this gives a curve as in Figure 1.2.

1.6 H_2^+: A PROTOTYPE MOLECULE

Within the Born–Oppenheimer approximation, we wish to solve

$$\hat{H}_{el}\Psi_{el} = E_{el}\Psi_{el} \tag{1.18}$$

where

$$\hat{H}_{el} = -\frac{\hbar^2}{2m}\nabla^2 - \frac{e^2}{4\pi\varepsilon_0|\mathbf{R}_1 - \mathbf{r}|} - \frac{e^2}{4\pi\varepsilon_0|\mathbf{R}_2 - \mathbf{r}|} \tag{1.19}$$

You are referred to any standard quantum chemistry text for the details of the solution of the problem. Equation (1.18) is soluble numerically, and the best solution is that of Wind (1965) who found an electronic energy of $-0.6026342\ E_h$* at $R_e = 2.00\ a_0$. In Table 1.1 we record several binding energies and R_e values. Very little direct experimental data is available for H_2^+, since the ground electronic state is the only one which is bound.

Figure 1.3 is a contour diagram for the lowest energy wavefunction (actually the square of the wavefunction, as discussed in a later chapter). Of particular interest is the fact that the wavefunction around each nucleus looks roughly spherical: in the region of a

Table 1.1 Dissociation energy D_e and equilibrium bond distances of H_2^+.

	D_e/E_h	R_e/a_0
Best numerical	0.1026342	2.003
Simple LCAO	0.0647	2.494
LCAO best exponent	0.0827	2.003
LCAO + $2p_\sigma$	0.0996	2.003
Elliptic orbitals	0.1019	2.003

* See Section 1.9 for a discussion of 'atomic units'.

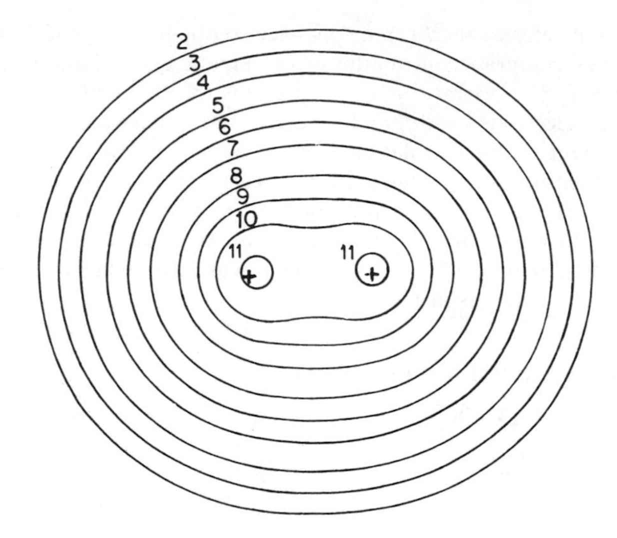

Figure 1.3 Contour diagram for $1\sigma_g$ orbital of H_2^+ (the plotted quantity is the square of the wavefunction)

nucleus, the electron will see a slightly distorted hydrogen atom, and we might be tempted to write the *molecular* orbital as contributions from *atomic* orbitals centred on the constituent atoms. This is referred to as the linear combination of atomic orbitals (LCAO) technique.

Suppose that $\phi_A(\mathbf{r})$ and $\phi_B(\mathbf{r})$ represent 1*s* atomic orbitals centred on nuclei A, B. We would write the molecular orbital

$$= C_A\phi_A + C_B\phi_B \tag{1.20}$$

and it is clear that $C_A = \pm C_B$ by symmetry. The positive combination gives a contour diagram very similar to Figure 1.3. It is obviously desirable to have a criterion for determining how well this LCAO wavefunction reproduces the exact solution, and we could also ask whether the 'form' of each atomic orbital in the molecule is different from the free atom; there is some evidence that the atomic orbitals 'contract' on molecule formation, as we will see shortly.

In the case of a polyatomic molecule, we would seek to write

$$\psi_{MO} = \Sigma\, C_k\phi_k$$

where the ϕ_k are centred on the various nuclei, and we obviously need a systematic procedure for finding the LCAO expansion coefficients C_k.

1.7 THE VARIATION PRINCIPLE

The equation $x^2 - 4x + 4 = 0$ has an exact algebraic solution. The equation $x\exp(x) = 1$ does *not*, but that does not prevent our finding a solution of the latter equation by numerical techniques. Any three-body problem, where three bodies interact with each other, is insoluble algebraically. It is irrelevant whether the force is electromagnetic, gravitational or whatever. We seek a solution of

$$\hat{H}\Psi = E\Psi \tag{1.21}$$

For the sake of argument, let us go through the following steps. Multiply from the left by Ψ^* and integrate over space:

$$\int\Psi^*\hat{H}\Psi\mathrm{d}\tau = E\int|\Psi|^2\mathrm{d}\tau \tag{1.22}$$

We could therefore in principle calculate the eigenvalue from

$$E = \int\Psi^*\hat{H}\Psi\mathrm{d}\tau/\int|\Psi|^2\,\mathrm{d}\tau \tag{1.23}$$

There is at first sight little to be gained from such a calculation, because an extra integration is involved. In the general case, however, we will never be able to solve the eigenvalue equation exactly, so our approximate wavefunction will *never* satisfy $\hat{H}\Psi = E\Psi$.

> *Problem 1.3*
> Decide whether $\tilde{\Psi} = x(a - x)$ is an eigenfunction of the 'electron in a box' problem. (*Answer:* it isn't.)

The ratio $\hat{H}\Psi/\Psi$ will in fact vary with position in space, and historically this gave rise to the so-called local energy method.

It turns out that, for any approximate wavefunction $\tilde{\Psi}$ with the correct boundary conditions, if we calculate

$$\tilde{E} = \int\tilde{\Psi}^*\hat{H}\Psi\,\mathrm{d}\tau/\int|\tilde{\Psi}|^2\,\mathrm{d}\tau \tag{1.24}$$

we will always find that $\tilde{E} \geqslant E_{\text{true}}$, the 'true' energy, and this is known as the *variation principle*. This result is proved in all elementary quantum mechanics texts, and relies on the fact that the approximate wavefunction $\tilde{\Psi}$ can be expressed in terms of the exact eigenfunctions Ψ_k as

$$\tilde{\Psi} = \Sigma C_k\Psi_k$$

where $\hat{H}\Psi_k = E_k\Psi_k$. The variation principle guarantees that the energy calculated from equation (1.21) will always be higher than the 'true' energy, and this gives a criterion for the accuracy of the guess. The variation principle in this form only applies to the lowest energy state of each symmetry.

> *Problem 1.4*
> Calculate $\tilde{E}$ for the trial function $\tilde{\Psi} = x(a - x)$ and compare your answer with the true energy $h^2/8ma^2$. (*Answer:* the ratio is $10/\pi^2$.)

In the case of H_2^+, if we use the simple LCAO function $\Psi_{MO} = C_A\phi_A + C_B\phi_B$ with $\phi = \sqrt{1/a_0^3\pi}\exp(-r/a_0)$, we find the potential energy surface shown in Figure 1.2: it lies *above* the exact potential energy surface for all R. Table 1.1 shows that the equilibrium bond distance is fairly well represented, but the binding energy is seriously in error. We note that as R increases, the approximate solution tends quickly to the true one. As we will see in a later chapter, this is fortuitous and happens rarely in such calculations.

Best use is made of the variation principle by introducing some parameters $C_1, C_2, \ldots, C_n$ into the trial wavefunction and then ensuring that these have their *optimum* values when $\partial E/\partial C_i = 0$. Thus for the simple LCAO treatment of H_2^+, we could have written $\phi = \sqrt{\zeta^3/\pi a_0^3}\exp(-\zeta r/a_0)$ and varied the *orbital exponent*. Finkelstein and Horowitz (1928) showed that this does indeed lead to a better binding energy (Table 1.1). The best value of ζ is 1.1, which is usually interpreted by saying that the H atom 1*s* orbital contracts on molecule formation.

The best H_2^+ variational calculations are those of James (1935), who used elliptic orbitals of the form $\phi = \exp(-\delta\mu)(1 + C\nu^2)$, where $\mu = (r_1 + r_2)/R$, $\nu = (r_1 - r_2)/R$ and δ, C are two variational parameters.

The difficulty in introducing a nonlinear parameter into the wavefunction is that each energy calculation is unique. We have to struggle with an integration and then find a minimum. No two problems ever look alike.

In the *linear variation method* we choose a set of *fixed* functions $\phi_1, \ldots, \phi_n$ and express our approximate wavefunction as a linear combination.

$$\tilde{\Psi} = C_1\phi_1 + C_2\phi_2 + \ldots + C_n\phi_n \tag{1.25}$$

The functions ϕ_i may not vary through the calculation, only the numerical coefficients. Once again, it is demonstrated in all the elementary quantum mechanics texts that the energy is minimized when the following matrix equation is satisfied:

$$\mathbf{HC} = \tilde{E}\mathbf{SC} \tag{1.26}$$

The $n \times n$ *Hamiltonian matrix* **H** has elements

$$(H)_{ij} = \int \phi_i^* \hat{H} \phi_j \mathrm{d}\tau$$

and the *overlap matrix* **S** has elements

$$(S)_{ij} = \int \phi_i^* \phi_j \mathrm{d}\tau$$

Equation (1.26) is a generalized matrix eigenvalue equation. For each value of i the energy E_i gives an approximation to the ith state of the molecular system, and the column $\mathbf{C}_i$ comprises n coefficients which describe state i as a linear combination of the ϕ_k. The matrix eigenvalue problem is well known and well studied in science, engineering and many branches of pure and applied mathematics.

It turns out that each value of $\tilde{E}_i$ calculated from (1.26) is greater than or equal to the true energy of each state, so the linear variation method has the added bonus of giving approximations to each of n states simultaneously. This is MacDonald's theorem (1933).

Figure 1.4 illustrates the effect of Macdonald's theorem; as we add more ϕ's to the expansion (1.25) each approximate $\tilde{E}_i$ gets closer to the true $E_{\text{true},i}$ but $\tilde{E}_i \geqslant E_{\text{true},i}$.

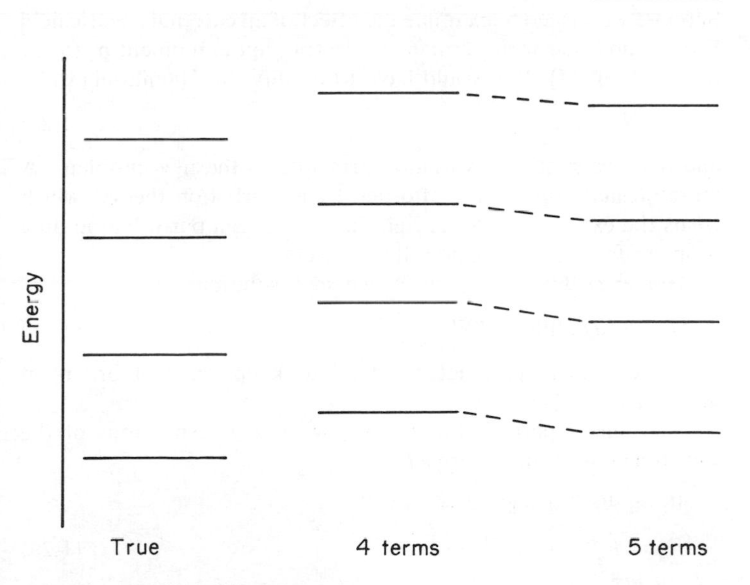

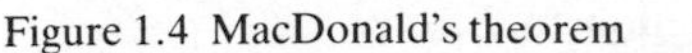
Figure 1.4 MacDonald's theorem

Thus, returning to the hydrogen molecule ion, we could express the MO as a linear combination of hydrogenic $1s$, $2s$, $2p_\sigma, \ldots$ *atomic* orbitals. Inclusion of $2\sigma p$ orbitals does indeed improve the bond length and the dissociation energy (Dickinson, 1933); see Table 1.1.

Problem 1.5

For the 'electron in a box' problem, take two functions $\phi_1 = x(a - x)^2$ and $\phi_2 = x^2(a - x)$. Both are plausible because they have the correct boundary conditions. Perform a linear variation calculation. You will find that the **H** and **S** matrices are

$$\mathbf{H} = -\frac{h^2 a^5}{8\pi^2 m}\begin{pmatrix} \frac{1}{15} & \frac{1}{60} \\ \frac{1}{60} & \frac{1}{5} \end{pmatrix} \qquad \mathbf{S} = a^7\begin{pmatrix} \frac{1}{105} & \frac{1}{140} \\ \frac{1}{140} & \frac{1}{105} \end{pmatrix}$$

Hence solve the eigenvalue problem.

1.8 PERTURBATION THEORY

Suppose we wished to examine the effect of an external electric field **E** on a molecule with permanent electric dipole moment $\mathbf{p}_e$ (as in the Stark effect). We would have to modify the Hamiltonian $\hat{H}_0$:

$$\hat{H} = \hat{H}_0 - \hat{\mathbf{p}}_e \cdot \mathbf{E} \tag{1.27}$$

and perhaps apply the variation principle to the new problem. A more elegant approach is afforded by perturbation theory, which treats the extra term on the right-hand side as a perturbation on a problem for which we know the answers.

Thus if $\hat{H}_0\Psi^{(0)}{}_k = E_k{}^{(0)}\Psi_k{}^{(0)}$ we seek solutions of

$$(H_0 + \lambda H^{(1)})\Psi = E\Psi \tag{1.28}$$

where λ is some parameter, added to keep track of orders of magnitude.

The aim of perturbation theory is to seek expansions of the wavefunction Ψ and energy E:

$$\begin{aligned} \Psi_k &= \Psi_k{}^{(0)} + \lambda\Psi_k{}^{(1)} + \ldots \\ E_k &= E_k{}^{(0)} + \lambda E_k{}^{(1)} + \ldots \end{aligned} \tag{1.29}$$

where $E^{(1)}$ is the first-order correction to E, etc.

Once again we refer you to almost any standard quantum chemistry text for a solution of the problem. For ease of notation, we assume that the Ψ_m are all *real*. The Rayleigh–Schrödinger method gives the following results: $\Psi_k^{(0)}$ is the state of interest in the absence of the perturbation and we assume that the $\Psi_k^{(0)}$ are orthogonal ($\int \Psi_k^{(0)}\Psi_l^{(0)}\mathrm{d}\tau = 0$) and normalized ($\int |\Psi_k^{(0)}|^2\mathrm{d}\tau = 1$).

$$E_k^{(1)} = \int \Psi_k^{(0)}\hat{H}^{(1)}\Psi_k^{(0)}\mathrm{d}\tau$$

$$E_k^{(2)} = -\sum_{m\neq k} \frac{\{\int \Psi_k^{(0)}\hat{H}^{(1)}\Psi_m^{(0)}\mathrm{d}\tau\}^2}{E_m - E_k} \qquad (1.30)$$

with corresponding results for $\Psi^{(1)}$ and $\Psi^{(2)}$, etc.

We will make use of the standard results of perturbation theory in subsequent chapters.

1.9 'ATOMIC UNITS'

Powers of 10 tend to be an embarassment in molecular structure calculations and it is usual in quantum chemistry to work with a set of units called 'atomic units', where the aim is to make the unnecessarily large or small powers of 10 disappear. These are shown in Table 1.2. The atomic unit of length a_0 is equal to the radius of the first Bohr orbit (and is usually called the bohr), whilst the atomic unit of energy, the hartree, is twice the energy of a ground-state hydrogen atom.

Table 1.2 Atomic units.

Quantity	Symbol	Value
length	a_0 (bohr)	5.2918×10^{-11} m
mass	m_e	9.1095×10^{-31} kg
time	t	2.4189×10^{-17} s
energy	E_h (hartree)	4.3598×10^{-18} J
charge	e	1.6022×10^{-19} C
angular momentum	$k = \hbar/2$	1.0546×10^{-34} J s
electric field	E	5.1423×10^{-11} V m^{-1}
electric field gradient	$-V_{zz}$	9.7174×10^{21} V m^{-2}
magnetic induction	B	2.3505×10^{5} T
electric dipole	p_e	8.4784×10^{-30} C m
electric quadrupole	Θ_ε	4.4866×10^{-40} C m^2
magnetic moment	p_M	1.8548×10^{-23} J T^{-1}
polarizability	α	1.6488×10^{-41} C^2 m^2 J^{-1}
magnetizability	$\varkappa$, χ	7.8910×10^{-29} J T^{-2}

It is also conventional to rewrite the basic equations of quantum chemistry in dimensionless form; for example, a 'reduced' length r' is given by r/a_0, etc., and you should readily verify that, for example, a H-atom Schrödinger equation

$$\left\{-\frac{h^2}{8\pi^2 m}\nabla^2 - \frac{e^2}{4\pi\varepsilon_0 r}\right\}\Psi(\mathbf{r}) = E\Psi(\mathbf{r}) \tag{1.31}$$

can be rewritten

$$\left\{-\frac{1}{2}\nabla'^2 - \frac{1}{r'}\right\}\Psi(\mathbf{r}') = E'\Psi(\mathbf{r}') \tag{1.32}$$

where the r' and E' are now just *numbers*, r/a_0 and E/E_h respectively. We usually drop the primes and refer to 'atomic units'.

CHAPTER 2

Self-Consistent Fields

2.1 THE PAULI PRINCIPLE

We discussed solution of the time-independent Schrödinger equation for a one-electron atom in Chapter 1. Let us extend the treatment to a two-electron 'atom' consisting of a nucleus of charge *Ze* and two 'electrons' which do not repel each other. (The reason for making this rather drastic approximation will appear shortly.) Labelling the electrons 1 and 2, we wish to solve, in the notation of Chapter 1, the eigenvalue problem

$$((\hat{h}(\mathbf{r}_1) + \hat{h}(\mathbf{r}_2))\Psi(\mathbf{r}_1, \mathbf{r}_2) = E\Psi(\mathbf{r}_1, \mathbf{r}_2) \tag{2.1}$$

It can be easily solved using the standard 'separation of variables' technique: *assume* that

$$\Psi(\mathbf{r}_1, \mathbf{r}_2) = \phi(\mathbf{r}_1) \times \chi(\mathbf{r}_2) \tag{2.2}$$

substitute and we find

$$\frac{1}{\phi(\mathbf{r}_1)}\,\hat{h}(\mathbf{r}_1)\phi(\mathbf{r}_1) + \frac{1}{\chi(\mathbf{r}_2)}\,\hat{h}(\mathbf{r}_2)\chi(\mathbf{r}_2) = E \tag{2.3}$$

Each term on the left-hand side is independently variable, so each must individually equal a constant, i.e.

$$\hat{h}(\mathbf{r}_1)\phi(\mathbf{r}_1) = E_1\phi(\mathbf{r}_1); \qquad \hat{h}(\mathbf{r}_2)\chi(\mathbf{r}_2) = E_2\chi(\mathbf{r}_2) \tag{2.4}$$

and so we recover an *orbital* picture. The many-electron wavefunction is given as a product of orbitals.

Clearly, we should not have ignored the electron–electron repulsion. It is *not* negligible and it is not constant. Inclusion of the $e^2/4\pi\varepsilon_0|\mathbf{r}_1 - \mathbf{r}_2|$, however, means that the total wavefunction *cannot* be written as a product of orbitals. We will see shortly how to calculate the best possible orbitals for a many-electron system, if we wish to retain the orbital picture of molecular structure.

Let us return to the two-electron atom problem and for ease of notation we label the two lowest energy orbitals $1s$, $2s$. Ignoring electron spin, the first three product wavefunctions for our (hypothetical) atom are

$$\begin{aligned} \Psi_{11}(\mathbf{r}_1, \mathbf{r}_2) &= 1s(\mathbf{r}_1)1s(\mathbf{r}_2) \qquad E = 2E_{1s} \\ \Psi_{12}(\mathbf{r}_1, \mathbf{r}_2) &= 1s(\mathbf{r}_1)2s(\mathbf{r}_2) \\ \Psi_{21}(\mathbf{r}_1, \mathbf{r}_2) &= 2s(\mathbf{r}_1)1s(\mathbf{r}_2) \end{aligned} \Bigg\} \quad E = E_{1s} + E_{2s}$$

in a fairly obvious notation. We need to enquire whether these three wavefunctions are acceptable descriptions of spectroscopic states.

The only physical observable is the total electron density, and the first point to notice is that, since electrons are indistinguishable according to the Heisenberg principle, they must appear on an equal footing in the electron density. The first wavefunction satisfies this criterion, but neither the second nor the third do because in each case it appears that *one* of the electrons is being treated differently from the other. (Electron 1 is apparently different from electron 2, because electron 1 is 'in' a 1*s* orbital in Ψ_{12} whereas it is 'in' a 2*s* orbital in Ψ_{21}.)

The combinations

$$\Psi_{12} + \Psi_{21}; \qquad \Psi_{12} - \Psi_{21}$$

do, however, satisfy the criterion: the electron density is symmetric in electron names, and so either combination is a suitable building block for the description of spectroscopic states. We label a wavefunction as symmetric or antisymmetric according to whether it remains unchanged or changes sign under interchange of electron labels. Ψ_{11} and $(\Psi_{12} + \Psi_{21})$ are *symmetric* whilst $(\Psi_{12} - \Psi_{21})$ is *antisymmetric*.

Problem 2.1
Prove the last statement.

So far we have made no mention of electron spin. Electron wavefunctions can be written as a product of a spatial part and a spin part, and the same considerations about indistinguishability have to apply to the square of the spin wavefunction. If the four spin states of electron 1 and 2 are $\alpha(s_1)$, $\beta(s_1)$, $\alpha(s_2)$, $\beta(s_2)$ then allowed combinations are obviously

$$\left.\begin{array}{c}\alpha(s_1)\alpha(s_2)\\ \alpha(s_1)\beta(s_2)+\beta(s_2)\alpha(s_1)\\ \beta(s_1)\beta(s_2)\end{array}\right\}\ \text{symmetric}$$

$$\alpha(s_1)\beta(s_2)-\alpha(s_2)\beta(s_1)\quad \text{antisymmetric}$$

and the *total* wavefunctions, which describe spectroscopic states, are formed by combining together the spatial functions with the spin functions.

It turns out as an experimental fact that there is a further requirement on the *total* wavefunction, usually referred to as the *generalized Pauli principle*, which states that any acceptable electronic wavefunction must be antisymmetric. So if we relabel electrons 1 and 2 as 2 and 1, the total electronic wavefunction must change sign. This means that we should only combine together *symmetric* spatial functions with *antisymmetric* spin functions and vice versa.

Acceptable electronic wavefunctions, capable of describing spectroscopic states, are therefore given by

$$\Psi_{11}(\alpha(s_1)\beta(s_2)-\alpha(s_2)\beta(s_1))$$

$$(\Psi_{12}+\Psi_{21})(\alpha(s_1)\beta(s_2)-\alpha(s_2)\beta(s_1))$$

$$(\Psi_{12}-\Psi_{21})\left\{\begin{array}{l}\alpha(s_1)\alpha(s_2)\\ \alpha(s_1)\beta(s_2)+\alpha(s_2)\beta(s_1)\\ \beta(s_1)\beta(s_2)\end{array}\right.$$

and are shown on an energy level diagram in Figure 2.1. Note that this diagram is a *spectroscopic state* diagram, not an orbital energy diagram. Because we have temporarily ignored electron repulsion, all four excited states are predicted to have the same energy. When electron repulsion is included in the calculation, this degeneracy is partially removed giving a singlet and a triplet excited states. The degeneracy of the triplet state can be removed in a Zeeman experiment, which we will discuss in some detail in a later chapter.

If we wish to extend this simple treatment to an n-electron system (where again electron repulsion is temporarily ignored), it proves convenient to work with *spin orbitals*, and we denote the composite space and spin variable $\mathbf{x} = \mathbf{r}s$.

Suppose we wish to describe a component of the quartet state of a three-electron atom arising from orbital configuration $1s^1\ 2s^1\ 3s^1$ with each electron having α spin. We could write

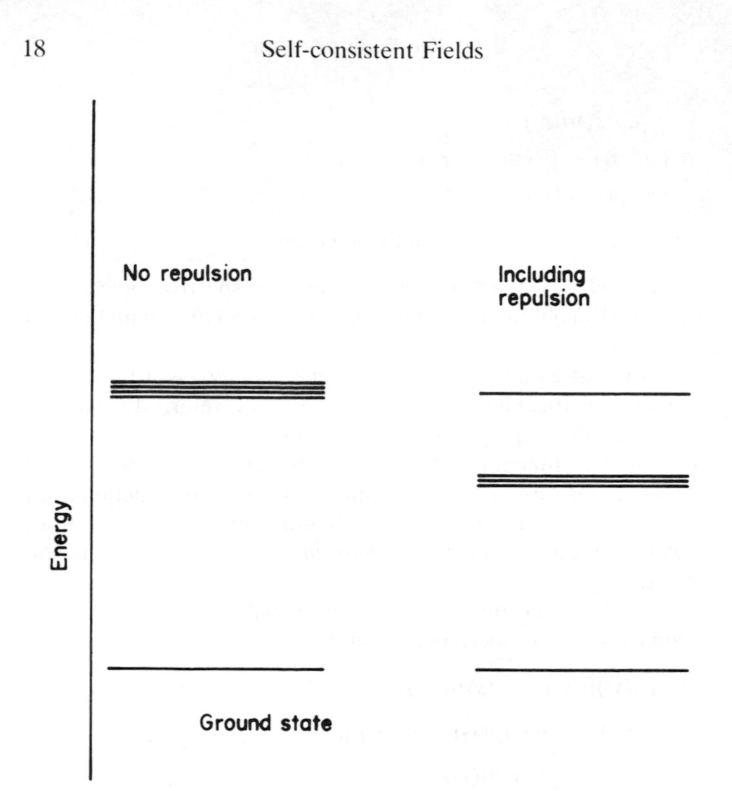

Figure 2.1 Schematic energy level diagram for a two-electron atom (a) ignoring electron repulsion, (b) including the effect of electron repulsion

$$1s\alpha(\mathbf{x}_1)\ 2s\alpha(\mathbf{x}_2)\ 3s\alpha(\mathbf{x}_3)$$

but by itself this could not represent a spectroscopic state because the indistinguishability criterion is not satisfied. Other contributions such as

$$-1s\alpha(\mathbf{x}_2)\ 2s\alpha(\mathbf{x}_1)\ 3s\alpha(\mathbf{x}_3)$$

would also be needed, and Slater (1929) noticed that the totality of all such permutations could be written as a 3 × 3 determinant (called a Slater determinant in his honour)

$$D = \begin{vmatrix} 1s\alpha(\mathbf{x}_1) & 1s\alpha(\mathbf{x}_2) & 1s\alpha(\mathbf{x}_3) \\ 2s\alpha(\mathbf{x}_1) & 2s\alpha(\mathbf{x}_2) & 2s\alpha(\mathbf{x}_3) \\ 3s\alpha(\mathbf{x}_1) & 3s\alpha(\mathbf{x}_2) & 3s\alpha(\mathbf{x}_3) \end{vmatrix} \tag{2.5}$$

This automatically satisfies the antisymmetry requirement, because interchanging (say) columns 1 and 2 of D changes its sign and corresponds physically to renaming electrons 1 and 2 as 2 and 1.

A Slater determinant is the smallest possible building block for a molecular electronic structure calculation.

2.2 THE EFFECT OF ELECTRON REPULSION

We discussed originally a pseudo 'two-electron atom', where the two electrons did not repel each other. Once electron repulsion is taken into account, the separation of variables technique does not work, and it proves impossible to factor the two-electron wavefunction into an orbital picture. This is not a failing of quantum chemistry; the same problem arises in *any* area of physics when the mutual interaction of three or more bodies has to be considered.

We will see in the next section that it proves possible to retain the orbital picture, but as an *approximate* picture, provided that we are prepared to follow a certain variational route to determine the best form of these orbitals.

2.3 SELF-CONSISTENT FIELDS (SCF)

The basic physical idea of the SCF method is that each electron moves in an *average* field due to the nuclei and remaining electrons, and so the effect of electron repulsion is formally included. Figure 2.2 illustrates a two-electron atom whose electrons are labelled 1 and 2. Electron 1 occupies orbital ϕ_1 and corresponds to an electron density $-e\phi_1^2$. The potential seen by electron 2 is therefore

$$V = -\frac{Ze^2}{4\pi\varepsilon_0|\mathbf{R} - \mathbf{r}_2|} + \frac{e^2}{4\pi\varepsilon_0}\int \frac{\phi_1^2(\mathbf{r}_1)}{|\mathbf{r}_1 - \mathbf{r}_2|}\,\mathrm{d}\tau_1 \tag{2.6}$$

and we can write formally a one-electron eigenvalue equation

$$\left\{-\frac{\hbar^2}{2m}\nabla_2^2 + V\right\}\phi_2 = E\phi_2 \tag{2.7}$$

for electron 2. Obviously we need to know ϕ_1 to calculate ϕ_2 etc., and we can imagine that some kind of iterative calculation will be necessary in order to calculate both ϕ_1 and ϕ_2.

We have, however, ignored spin and antisymmetry, and the actual *form* of the Hartree–Fock operator is more complicated than given above. Nevertheless, it proves possible to write an eigenvalue equation

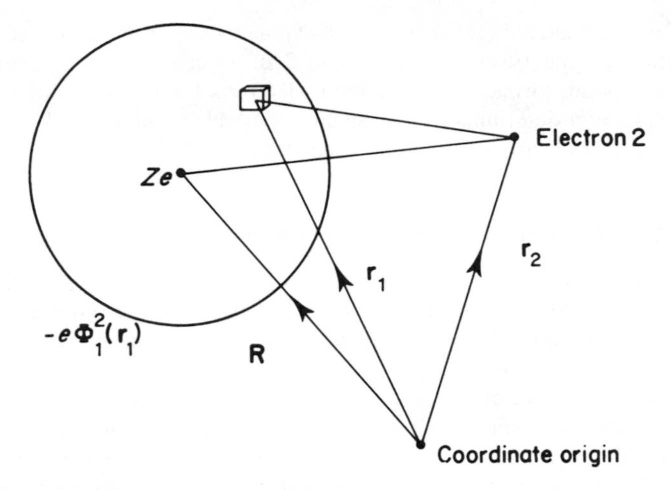

Figure 2.2 Elementary construct needed to discuss self-consistent fields. Electron 1 is described by orbital ϕ_1

$$\hat{h}^{\mathrm{F}}\phi_i = e_i\phi_i \tag{2.8}$$

which incorporates the variation principle. The *form* of all the orbitals is varied simultaneously in order to attain the lowest possible energy.

The precise details of the Hartree–Fock operator need not concern us; we note, however, that it is *defined* in terms of the electron density (the ϕ_i^2's) and so an *iterative* solution of the eigenvalue equation is necessary.

The e_i are called *orbital energies*. The Hartree method owes its name to W. Hartree and dates from 1927. Fock incorporated antisymmetry into the method in 1930, and all early applications were to atoms. Because of the high symmetry of an atom, the differential equation can be solved numerically. D.R. Hartree's book *The Calculation of Atomic Structures* was published in 1957 and reflected the state of the art at that time.

For molecular calculations, numerical integration of the Hartree–Fock equations proves impossible, and Roothaan introduced an LCAO version of SCF theory in 1951. Each SCF orbital is written as a linear combination of fixed orbitals $\phi_1, \phi_2, \ldots, \phi_n$ ('fixed' in the sense that they are not allowed to vary as the SCF calculation proceeds). These orbitals are usually called 'basis functions', and from n basis functions we will generate n LCAO functions, so that

$$\psi_k(\mathbf{r}) = \sum_{i=1}^{n} C_{k,i}\phi_i(\mathbf{r}) \tag{2.9}$$

and for each value of k we can collect the LCAO coefficients into a column vector $\mathbf{C}_k$. Each vector $\mathbf{C}_k$ is found by solving the matrix Hartree–Fock equations

$$\mathbf{h}^{\mathrm{F}}\mathbf{C}_k = E_k\mathbf{S}\mathbf{C}_k \tag{2.10}$$

where **S** is the matrix of overlap integrals,

$$S_{ij} = \int \phi_i^*(\mathbf{r})\phi_j(\mathbf{r})\mathrm{d}\tau$$

The calculation of $\hat{h}^{\mathrm{F}}$ involves the evaluation of one electron integrals $\int \phi_i^*(\mathbf{r}_1)\hat{h}(1)\phi_j(\mathbf{r}_1)\mathrm{d}\tau_1$ and a large number of two-electron integrals $\int \phi_i^*(\mathbf{r}_1)\phi_j^*(\mathbf{r}_2)\hat{g}(1, 2)\phi_k(\mathbf{r}_1)\phi_l(\mathbf{r}_2)\mathrm{d}\tau_1\,\mathrm{d}\tau_2$ in the notation of Chapter 1. The former integrals give contributions to the kinetic energy of the electron–nuclear attraction, the latter integrals give a contribution to the energy from the electron repulsions.

For a closed-shell molecule with M electron pairs, the lowest energy orbitals $C_1 \ldots C_M$ are occupied, and we refer to the remaining $n - M$ as 'virtual' orbitals. They have the physical interpretation of describing the potential seen by a 'test charge' in the presence of the electron density due to the neutral molecule.

The LCAO orbitals are usually chosen so that they are normalized and orthogonal,

$$\int \psi_k^*(\mathbf{r})\psi_l(\mathbf{r})\mathrm{d}\tau = \delta_{kl} \tag{2.11}$$

and this can be written in terms of the columns as

$$\mathbf{C}_k{}^+\mathbf{S}\mathbf{C}_l = \delta_{kl} \tag{2.12}$$

where $\mathbf{C}^+$ is the Hermitian transpose of **C**.

Again, we remark that h^{F} is *defined* in terms of the one- and two-electron integrals, and the C's, so an iterative solution of the eigenvalue problem is usual. Because the basis functions $\phi_1, \ldots, \phi_n$ do not change during the iterative cycles, it is usual to calculate all the integrals at the start of the SCF calculation, and store them.

Historically, the 'two-electron integrals problem' was the greatest problem in quantum chemistry; let us see why. Firstly, we known from atomic SCF calculations that the 'best' atomic orbitals to use in molecular calculations are those with an exponential factor $\exp(-\zeta r)$: such orbitals are referred to as Slater-type orbitals (STO), and a great deal of effort has been expended on calculating variationally the 'best' exponents (see Clementi and Raimondi, 1963).

Two-electron integrals are numerous: if we perform an SCF calculation with n basis functions, we will have to calculate $p = n(n + 1)/2$ of each type of one-electron integral, and at most $q = p(p + 1)/2$ two-electron integrals. So, if $n = 150$, we need to calculate, store and manipulate 7.4×10^7 two-electron integrals. The integrals are six-dimensional integrals with a singularity (as $r_1 \to r_2$), and if each atomic orbital is centred on a different nucleus the integrals are essentially impossible to evaluate either numerically or analytically.

The great breakthrough came in the 1960s with the independent proposal of *Gaussian* orbitals by Boys and by McWeeny. A Gaussian orbital has exponential part $\exp(-\alpha r^2)$ and these orbitals have the great advantage that a product of two Gaussians can be expressed as single Gaussian, located along the line of centres of the two in the product. This means that an 'impossible' four-centre integral can be reduced to a fairly trivial one-centre problem.

It turns out to be convenient to work with *Cartesian Gaussians*; these have the form

$$x^l y^m z^n \exp(-r^2)$$

and we label an orbital of type $\exp(-r^2)$ an s-orbital, whilst the three orbitals $x\exp(-r^2)$, $y\exp(-r^2)$ and $z\exp(-r^2)$ are labelled p-orbitals. The *six* $x^2\exp(-r^2)$, $y^2\exp(-r^2)$, $z^2\exp(-r^2), \ldots$, $yz\exp(-r^2)$ are usually referred to as d-orbitals but strictly the combination $(x^2 + y^2 + z^2)\exp(-r^2)$ is an s-orbital. Some computer packages for SCF calculations eliminate this so-called 'spurious' d-function, but there is no particular reason for doing so.

Gaussian orbitals are widely used in molecular structure calculations, but they have serious defects. They fall off too rapidly with r, and hence more Gaussians have to be used than would be the case with Slater orbitals. Gaussians have the wrong behaviour at $r = 0$ and this incorrect behaviour can have serious consequences for properties which depend on a density at nuclear positions: for example, electron spin resonance coupling constants.

A very large number of Gaussian atomic orbital basis sets are available, and many software packages store a selection internally. A recent compilation has been given by Poirer, Kari and Czismadia (1985).

Basis sets are usually classified as follows. The simplest type is the *minimal basis set* where each atom is represented by a single orbital of each type as in descriptive organic chemistry. Thus a carbon atom would need 2 s-orbitals and 1 p-orbital of each type.

The basis functions would normally be a sum of primitive Gaussians. The basis set of Table (2.1) is a minimal basis set.

A *double zeta* basis set is a basis set comprising exactly double the number of functions in a minimal set. Each '1*s*' orbital is therefore represented by two basis functions, and so on.

An *extended* basis set is a generic name for anything more sophisticated than a minimal set.

A *polarization function* is an atomic orbital with angular momentum quantum number higher than the maximum necessary to describe the ground state of the neutral atom. Thus a *d*-orbital for carbon is a polarization function. Polarization functions are needed:

(a) in order to describe accurately the electron density in a molecule, where the symmetry is much lower than in an atom, and
(b) to describe the response of the electron density to an external field.

They should be routinely added, if possible, for molecular structure calculations. The optimum exponents for polarization functions have to be determined from *molecular* calculations.

There have been two philosophies for the construction of Gaussian basis sets. The first philosophy is to start at the descriptive organic chemistry level of, for example, a carbon atom being de-

Table 2.1 An STO/4G basis set for carbon. The 2*s* and 2*p* basis functions share the same set of primitives for reasons of programming efficiency.

Type	Exponent	Contraction	Primitive	Basis fn
s	167.716	0.05678	1	1
s	30.6899	0.26014	2	
s	8.5260	0.53285	3	
s	2.8297	0.29163	4	
s	6.8739	−0.06221	5	2
s	1.4880	0.00003	6	
s	0.4838	0.55886	7	
s	0.1858	0.49777	8	
p	6.8739	0.04368	5	3
p	1.4880	0.28638	6	
p	0.4838	0.58358	7	
p	0.1858	0.24631	8	

scribed by $1s$, $2s$ and $2p$ orbitals. One knows from *atomic* SCF calculations that Slater-type orbitals (STO) give the best description of electronic structure, so we start by performing a minimal basis set atomic SCF calculation and carefully optimizing the orbital exponents. The optimum exponents turn out to be 5.6727, 1.6083 and 1.5679 (Clementi and Raimondi, 1963). Note that the $2s$ and the $2p$ orbitals have slightly different orbital exponents. We then fit by least-squares techniques each STO to a linear combination of Gaussian-type orbitals (GTO), and typically each STO is represented as the fixed linear combination of GTO shown in Table 2.1. We refer to the basis set of Table 2.1 as an STO/4G basis set; as far as the SCF calculations are concerned, we are working with just five carbon basis functions.

If we wish to add a little more flexibility to the basis set, we can let the 'inner' and 'outer' portions of the valence orbitals vary independently in the molecular calculation, and we speak typically of an STO/4-31G basis set. This means that the $1s$ orbital is represented by four Gaussians, whilst the $2s$ and $2p$ valence orbitals are each represented by one orbital which is a combination of *three* Gaussians and one which is a diffuse orbital.

The second approach is to start with a set of Gaussian orbitals (usually referred to as *primitive* Gaussians) and to systematically optimize the exponents by repeatedly performing atomic SCF calculations. Thus for example, Table 2.2 shows a typical carbon Gaussian basis set. The orbital exponents were determined variationally by minimizing the SCF atomic energy, giving finally the 14 different orbital exponents.

On close investigation it turns out in the case of carbon that the first 6 primitives appear from molecule to molecule in roughly the same ratios as given in the column labelled 'contraction'; this linear combination of six primitives gives a representation of what the organic chemist would call the carbon $1s$ orbital'. Similar comments apply for primitives 7 through 9 and 11 through 13. Primitive 10 and primitive 14 have low exponents and are therefore diffuse. They are needed in order to compensate for the poor representation of the outer regions inherent in a Gaussian basis set. In practice therefore we group together primitives 1 through 6, 7 through 9, and 11 through 13 and imagine each combination as representing a single *contracted basis function*, often just referred to as a basis function. This process of contraction means that the variation principle is not allowed to operate completely in the SCF calculation, because we have added an artificial constraint in keeping certain ratios of cer-

Table 2.2 A typical large Gaussian orbital basis set suitable for carbon in molecular calculations.

Type	Exponent	Contraction	Primitive	Basis function
s	3047.0	0.00183	1	1
s	457.4	0.01404	2	
s	103.9	0.06884	3	
s	29.21	0.23218	4	
s	9.287	0.46794	5	
s	3.164	0.36231	6	
s	7.868	−0.11933	7	2
s	1.881	−0.16085	8	
s	0.5442	1.14346	9	
s	0.1687		10	3
p	7.868	0.06900	11	4
p	1.881	0.31642	12	
p	0.5442	0.74431	13	
p	0.1560		14	5

tain MO coefficients constant. In practice we only *store* integrals over (contracted) basis functions.

The question of choosing a basis set for a given calculation is not a trivial one, and one has to consider several factors; not the least is that of *cost*. For small and medium-sized molecules the final choice is usually determined by whether one is only interested in a trend across a series of related molecules. Are excited states or Rydberg states important? Is the molecule negatively charged? etc. We will refer to choice of basis set throughout the text.

2.4 OPEN-SHELL ELECTRONIC STATES

More often than not, we are concerned with the totally symmetric singlet ground state of an atom or molecule, and a state that can be so represented as a single Slater determinant of doubly occupied orbitals is often called a 'closed-shell' state.

It is clear, however, that any atom or molecule with an odd number of electrons cannot have a closed-shell ground state, and unfortunately not every electronic state can be treated by SCF theory. There are two main techniques for treating the open-shell problem. Firstly, we can envisage a simple extension to closed-shell SCF theory, where some of the orbitals are singly occupied with all spins parallel and we generally refer to such calculations as 'restricted Hartree–Fock' (RHF) calculations. It does prove possible to treat certain more general electronic states of atoms and molecules by the RHF technique.

The 'unrestricted Hartree-Fock' (UHF) method permits the α and β spin orbitals to vary *independently*. For a Li atom, we would take *three* spin orbitals $1s\alpha$, $1s'\beta$, $2s\alpha$ and allow the spatial part of all three to vary independently. The resultant orbitals are often said to be 'spin polarized'; in the Li case the outer $2s$ electron would spin polarize the inner shell, producing two slightly different spatial orbitals $1s$, $1s'$. Unfortunately, UHF wavefunctions are not spin eigenfunctions and they are therefore unsuitable in principle for representing spectroscopic states. Formally, the UHF wavefunction Ψ_{UHF} can be written as a sum of different spin states

$$\Psi_{UHF} = C_{2s+1}\Psi_{2s+1} + C_{2s+3}\Psi_{2s+3} + \ldots$$

where $s = \frac{1}{2}(n_\alpha - n_\beta)$ with n_α, n_β being the number of α-spin and β-spin electrons. Thus if $n_\beta = n_\alpha + 1$, the UHF wavefunction will contain contributions from a doublet, a quartet, etc., and the highest contribution will have multiplicity $n_\alpha + n_\beta + 1$.

It often happens that $C_{2s+1} \approx 1$, and in this case the major 'contaminating' spin state is usually Ψ_{2s+3}. In such circumstances, the easiest correction is to remove the contribution from Ψ_{2s+3} only.

Despite the comment regarding spin eigenfunctions, the UHF technique is widely used for investigating open-shell species.

CHAPTER 3

Software

Over the last 20 years, international collaboration and cooperation on a scale rarely witnessed in science has led to the development of several very sophisticated software packages for *ab initio* molecular electronic structure calculations. Many of these packages are essentially public domain software available for a nominal fee from their authors or from the Quantum Chemistry Program Exchange. Most academic, and a growing number of industrial, computer installations have these packages available.

We will generally refer in the text to two such packages, GAUSSIAN 86 and GAMESS. GAUSSIAN 86 is a logical extension of the GAUSSIAN XY series (where XY = year of issue) of packages associated with Hehre, Ditchfield, Radom and Pople. GAMESS (General Atomic and Molecular Electronic Structure System) is a development of an earlier package called HONDO, which in turn owes some of its parentage to the early GAUSSIAN 70. Input is generally user-friendly, simple and common to many packages.

At its very simplest, an *ab initio* package consists of programs to calculate one- and two-electron integrals over Gaussian orbitals and to perform SCF calculations for closed- and open-shell molecules. We could also add to this minimal list programs for the calculation of properties such as electric dipole moments, and programs to aid with analysis of the wavefunction.

Some packages such as POLYATOM make extensive use of molecular symmetry in order to reduce the number of integral evaluations, and many packages store internally commonly used Gaussian basis sets. At its simplest, a package will typically consist of 20000 lines of FORTRAN.

More modern packages contain procedures to calculate molecular geometries, force fields, molecular properties such as electric dipole polarizability and magnetizability which measure the response of a molecule to an external electric or magnetic field, and have the ability to deal with the electron correlation. Such packages are therefore suitable both for highly accurate calculations on small molecules, and for calculations on large molecules of pharmaceutical interest.

3.1 BASIS SETS

To take a concrete example, GAMESS has the following basis sets stored internally:

(a) STO/nG: The STO/nG basis sets for first and second row atoms due to Collins, Schleyer, Binkley and Pople (1976).
(b) SV: The split valence basis sets due to Dunning and Hay.
(c) SV 3-21G: The split valence basis sets STO/3-21G, STO/4-31G and STO/6-31G due to Frisch, Pople and Binkley (1984).
(d) TZV: The triple-zeta basis sets due to Dunning (1971) and to McClean and Chandler (1980) for first-row, second-row and first-transition row elements.
(e) TZVP: The TZV basis augmented with polarization functions.

It is also possible to add basis functions or to input a general basis set.

3.2 MOLECULAR GEOMETRY

The majority of molecular structure calculations are performed within the Born–Oppenheimer approximation, which means that the nuclei are clamped in position for the purpose of calculating the *electronic* wavefunction. The nuclei are then taken to move in a potential provided by the electrons. We therefore need to specify a molecular geometry, which means that we need to specify bond lengths and bond angles. Obviously, we can use the variation principle to *predict* a molecular geometry, and we will return to this theme in a later chapter.

Figure 3.1 shows ethene. We have looked up the geometry in a standard tabulation and chosen coordinate axes. The nuclear Cartesian coordinates are given in Table 3.1.

An alternative and chemically more appealing way to specify the geometry is to specify the atomic positions in terms of bond lengths, bond angles and dihedral angles and let the program sort out conversion to Cartesian coordinates. If we wish to let the package automatically optimize the geometry, this latter method is essential. Input is usually done via a so-called ZMATRIX as follows.

Each nucleus is numbered sequentially and specified on a single

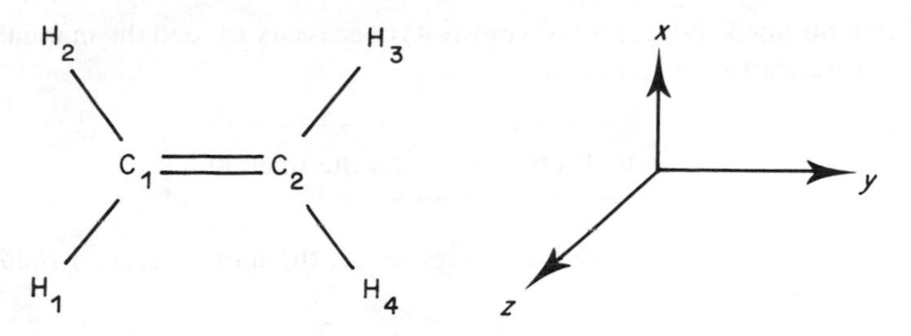

Figure 3.1 Numbering of atoms for ethene calculation and choice of coordinate axes

input record so that the *n*th record is used to specify the nature and location of this nucleus in terms of previously determined ones. Typically, record – consists of seven items which we can write

TAG, NUC1, R1, NUC2, A12, NUC3, A123.

TAG is used to give nucleus N a name by which it will be subsequently known. Typically it is the chemical symbol with an identifying number of letter.

NUC1 is a previously defined nucleus for which the internuclear distance will be given.

R1 is the distance between the two nuclei (which need not be bonded).

NUC2 is a second nucleus for which the bond angle N, NUC1, NUC2 is to be given.

A12 is the value of this angle.

NUC3 is a third nucleus for which a dihedral angle, N, NUC1, NUC2, NUC3 is to be given.

A123 is the angle between the planes (N, NUC1, NUC2) and (NUC1, NUC2, NUC3). The *sign* is determined from simple rules.

Most packages have somewhat more sophisticated facilities than

Table 3.1 Cartesian coordinates for ethene.

Atom	x/a_0	y/a_0	z/a_0
H_1	−1.753	−2.291	0.0
H_2	1.753	−2.291	0.0
C_1	0.0	−1.279	0.0
C_2	0.0	1.279	0.0
H_3	1.753	2.291	0.0
H_4	−1.753	2.291	0.0

that outlined above, and obviously it is necessary to read the manual before starting a calculation!

'When all else fails, read the manual'!

For ethene, numbered as in Figure 3.1, the input records would typically be:

```
C1
C2  C1  RCC
H1  C1  RCH  C2  120
H2  C2  RCH  C1  120  H1  0
H3  C1  RCH  C2  120  H2  180
H4  C2  RCH  C1  120  H3  0
```

where we have used RCC and RCH for the C—C bond and C—H bond distances respectively.

3.3 A CLOSED-SHELL CALCULATION

In the special case of GAMESS, input is controlled by *keywords*. Many of the package options (such as spin multiplicity and overall molecular charge) can be left to default settings, and typical input for a single SCF calculation on ethene using the STO/3G basis set would be:

TITLE
ETHENE SCF CALCULATION
ZMATRIX
C1
C2 C1 RCC
(etc.)
CONSTANTS
RCC 2.557
RCH 2.024
END
BASIS STO/3G

Keywords are printed bold.

Output from the calculation consists of a total energy (-77.069401 E_h in this case) together with a set of orbital energies and LCAO coefficients, as shown in Table 3.2. The output is self-evident: each basis function (1–14) consists of a fixed linear combination of three

Table 3.2 Orbital energies e_i and LCAO-MO coefficients for ground-level state ethene at the STO/3G level.

e_i/E_h	−11.018	−11.017	−0.976	−0.748	−0.609	−0.523	−0.473	−0.319
H_1s	.0048	.0050	−.1187	.2239	.2593	.2142	−.3442	0.
H_2s	.0048	.0050	−.1187	.2239	−.2593	.2142	.3442	0.
C_1s	−.7019	−.7014	.1767	−.1379	0.	.0158	0.	0.
C_1s	−.0202	−.0308	−.4696	.4171	0.	−.0336	0.	0.
C_2s	−.7019	.7014	.1767	.1379	0.	.0158	0.	0.
C_2s	−.0202	.0308	−.4696	−.4171	0.	−.0336	0.	0.
H_3s	.0048	−.0050	−.1187	−.2239	−.2593	.2142	−.3442	0.
H_4s	.0048	−.0050	−.1187	−.2239	.2593	.2142	.3442	0.
C_1x	0.	0.	0.	0.	−.3919	0.	.3931	0.
C_2x	0.	0.	0.	0.	−.3919	0.	−.3931	0.
C_1y	.0023	−.0042	−.1057	−.1948	0.	−.5062	0.	0.
C_2y	−.0023	−.0042	.1057	−.1948	0.	.5062	0.	0.
C_1z	0.	0.	0.	0.	0.	0.	0.	.6371
C_2z	0.	0.	0	0.	0.	0.	0.	.6371

primitive gaussians (STO/3G basis set), and for example molecular orbital number 8, which has orbital energy $-0.3188\ E_h$, consists of $0.6371\ C_1(z) + 0.6371\ C_2(z)$ basis functions.

3.4 OPEN-SHELL CALCULATIONS

The lowest-energy cation of ethene is formed by removing a π-electron. If we assume that the cation geometry is identical to the parent molecule, all that is necessary to perform an open-shell calculation is to specify:

MULTIPLICITY 2

CHARGE + 1

in the input. As we saw in Chapter 2, open-shall states can be treated using either the restricted (RHF) or unrestricted (UHF) Hartree–Fock procedures, so it is necessary to specify the type of SCF calculation.

Table 3.3 shows the result of a restricted Hartree–Fock calculation for the ethene cation, calculated at the same nuclear geometry as for the neutral ground state molecule in Table 3.2. Molecular orbitals 1 through 7 are doubly occupied, and orbital 8 is singly occupied. It is obviously important to have some method of comparing one LCAO-MO wavefunction with another; a straightforward comparison of LCAO coefficents is definitely not recommended, and we return to this theme in a later chapter. For the minute, we note that the total energy was $-76.785\ 430\ E_h$, which gives a predicted ionization energy of $0.284\ E_h$ (or 7.73 eV).

A corresponding unrestricted Hartree–Fock calculation gives a total energy of $-76.786\ 762\ E_h$, slightly lower than the RHF results as one would anticipate. The expectation value of S^2 (the spin operator) is, however, totally unacceptable at 1.128 (for a pure doublet state the value should be $1/2(1 + 1/2) = 3/4$).

3.5 CORRELATION ENERGY

Table 3.4 records the total energy (electronic + nuclear repulsion) calculated by the SCF method for the geometry of Section 3.4 using a variety of different Gaussian basis sets. We can rank the calculations in order of 'goodness' according to the variation principle: the lower the energy, the 'better'. Thus we note for example the dramatic improvement in moving along the series STO/nG, caused by improving the description of the carbon inner shell. We also note

Table 3.3 Open-shell SCF orbital energies e_i and LCAO-MO coefficients for ethene cation.

e_i/E_h	−11.463	−11.463	−1.278	−1.064	−0.960	−0.881	−0.810	−0.502
H_1s	−0.0047	.0049	−0.957	−1.958	.2134	.1923	.3055	0.
H_2s	−0.0047	.0049	−0.957	−1.958	.2134	.1923	−0.3055	0.
C_1s	0.7023	−.7017	.1178	.1445	0.	−.0006	0.	0.
C_1s	0.0187	−.0300	−.4832	−.4660	0.	.0331	0.	0.
C_2s	0.7023	.7017	.1778	.1445	0.	−0.006	0.	0.
C_2s	0.0187	.0300	−.4832	.4660	0.	.0331	0.	0.
H_3s	−0.0047	−.0049	−.0957	.1958	−.2134	.1923	.3055	0.
H_4s	−0.0047	−.0049	−.0957	.1958	.2134	.1923	−.3055	0.
C_1x	0.	0.	0.	0.	−.4404	0.	−.4508	0.
C_2x	0.	0.	0.	0.	−.4404	0.	.4508	0.
C_1y	−0.0021	−.0049	−.1295	.1995	0.	−.5209	0.	0.
C_2y	0.0021	−.0049	.1295	.1995	0.	.5209	0.	0.
C_1z	0.	0.	0.	0.	0.	0.	0.	.6371
C_2z	0.	0.	0.	0.	0.	0.	0.	.6371

Table 3.4 Total energy of ethene with different basis sets.

Basis set	No. of basis functions n	E/E_h	Time(s)
STO/3G	14	−77.069 401	2.43
STO/4G	14	−77.624 194	3.28
STO/6G	14	−77.824 357	6.91
SV	26	−78.011 089	11.92
SV/3-21G	26	−77.598 679	5.36
SV/4-31G	26	−77.919 840	6.26
SV/6-31G	26	−78.002 631	7.35
TZVP	64	−78.059 474	143.46

that eventually the gains in total energy become smaller with increase in sophistication of basis set and with cost.

The experimental value of the total energy is unknown; and in any case direct comparisons between calculated values and the experimental value would have to be made with some caution. The SCF calculations refer to a fixed geometry and a non-relativistic Hamiltonian.

Nevertheless, Figure 3.2 shows the relationship between the 'experimental' value and various SCF energies. Because in SCF calculations electrons are assumed to move in an average potential, the best SCF calculation that could possibly be made (i.e., the 'SCF limit') would still given an energy higher than the true one. The SCF limit can be reached readily in *atomic* SCF calculations, where the Hartree–Fock equations can be integrated numerically. For molecules, however, calculations at the Hartree–Fock limit are unattainable.

The difference between the 'experimental' and the Hartree–Fock limit energies is called the *correlation energy*, and it is typically a few per cent of the total energy. Perhaps the most important consequence is that SCF calculations cannot in principle treat phenomena such as dispersion forces, which depend on the *instantaneous* interactions between particles.

Of practical importance is the *cost* (i.e. the computer resource consumed) of an SCF calculation, the cost rises roughly as $n^{3.5}$ where n is the number of basis functions. Thus for example, the TZVP calculation costs some 60 times more than the STO/3G.

In the next sections, we describe briefly some of the options available for treating electron correlation. They all have one feature in common: they are very costly in computer time and resource.

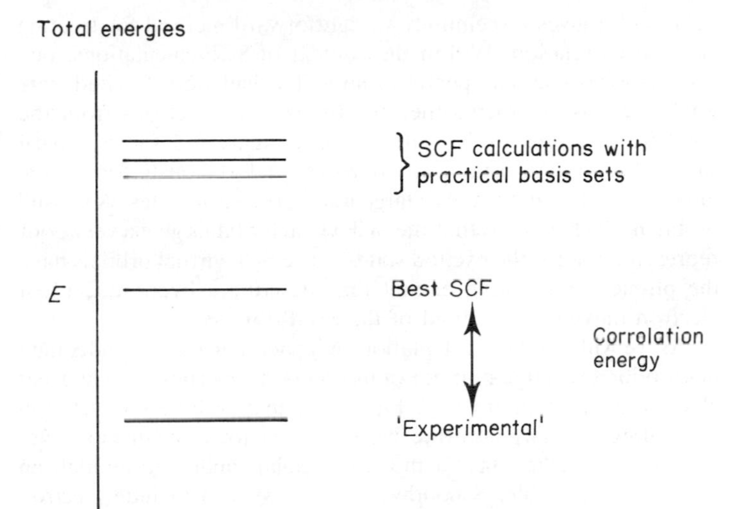

Figure 3.2 Comparison of SCF calculated energies and 'experimental' energies

3.5.1 Configuration interaction

Consider a very simple problem: we wish to treat electron correlation in a lithium atom. We perform an SCF calculation with a large basis set; the lowest energy orbital we label 1*s*, the next lowest 2*s*, etc., and we would write the ground state electronic configuration symbolically as $\Psi_0 = 1s^2 2s^1$. Electronic configurations such as $\Psi_1 = 1s^2 3s^1$, formed by exciting an electron from the 2*s* to the 3*s* (virtual) orbital, are called singly excited states; Ψ_0 and Ψ_1 would give a fair representation of those particular spectroscopic states.

Within the spirit of the variation principle, we know that it will be possible to improve *both* wavefunctions by writing

$$\Psi = C_0 \Psi_0 + C_1 \Psi_1$$

and solving the matrix eigenvalue problem

$$\mathbf{HC} = E\mathbf{SC}$$

to find the best values of C_0 and C_1. The lower energy solution will give us a better description of the ground electronic state, whilst the higher energy solution will similarly improve the description of the first excited state.

This process is generally referred to as configuration interaction

(CI), and it gives a seemingly straightforward method for treating electron correlation. Within the context of SCF calculations, one chooses a basis set and performs an SCF calculation. Excited state wavefunctions are then generated by exciting electrons from the filled SCF orbitals to the virtual ones. Figure 3.3 shows typical singly, doubly and triply excited states, and obviously for a large molecule there will be a very large number of such states. A second problem which arises is that the SCF virtual orbitals give a very poor representation for the excited states, since SCF virtual orbitals have the physical interpretation that they describe a hypothetical test electron moving in the field of the *neutral* molecule.

As we will see, SCF calculations in general give a very adequate description of a large number of molecular properties, so why treat electron correlation anyway? Firstly, we may well be interested in calculating a highly accurate wavefunction for a small molecule, in order to predict, say, a molecular polarizability to as high an accuracy as possible. Secondly, we may wish to include electron correlation simply because the property we are calculating is given poorly or not at all at SCF level: calculations of dispersion forces or of the dissociation products for the reaction $H_2 \rightarrow 2H$ at SCF level are automatically doomed to failure.

A detailed analysis of the CI process is outside the scope of this elementary text. We should say, however, that there are several logical steps to such a calculations. Firstly, one chooses a configuration state function (CSF) or perhaps several as required to describe the process under study. One then decides which excited states to take. Modern CI programs are written to permit single plus double excitations. It is then usually necessary to form the Hamiltonian matrix, which involves transforming the two-electron integrals from atomic orbitals to a sum over molecular orbitals. Finally one seeks

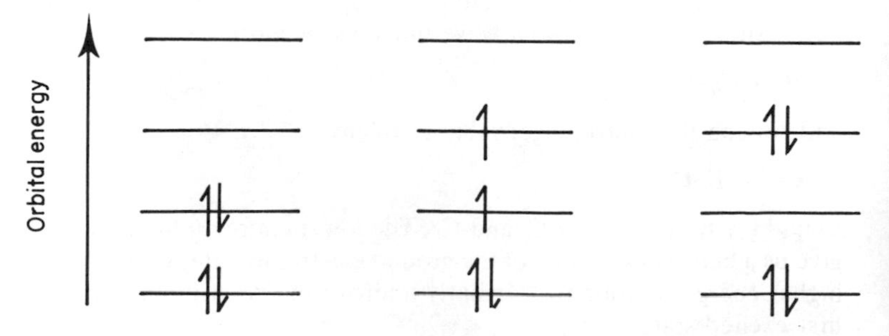

Figure 3.3 Closed-shell, singly- and doubly-excited configurations.

the lowest few eigenvectors of the Hamiltonian matrix. We should say that the selection of excited states can be done by using perturbation theory to examine the likely effect of a given excited state before including it in the CI expansion. For technical reasons, it turns out that there are certain advantages in performing CI calculations where *all* singly and doubly excited states within a chosen number of filled and virtual orbitals are included in the calculation. We refer to such calculations as CISD (CI single + double excitation), and we might for example choose to exclude all inner-shell orbitals and the high-energy virtual ones from a molecular calculation.

3.5.2 Multiconfiguration SCF (MCSCF)

The SCF-CI method works with a fixed set of molecular orbitals. We normally start from a small number of configurations (e.g. representing reactants and products in a chemical reaction), and generate a large number of excited state wavefunctions. The *form* of the individual molecular orbitals is unchanged throughout the calculation, and we noted earlier that this was a major problem in CI calculations, leading to a very slowly convergent expansion.

The MCSCF method starts from a linear combination of a small number of Slater determinants and seeks to optimize simultaneously the linear expansion coefficients and the (nonlinear) LCAO coefficients simultaneously. This is a more ambitious problem, and very few molecular applications have appeared in the literature to date.

3.5.3 Generalized valence bond (GVB)

Most elementary quantum chemistry texts give the simple MO and valence bond (VB) treatments of H_2. If we label the nuclei A, B and write ϕ_A, ϕ_B for atomic 1*s*-like orbitals centred on A and B, then the simple VB treatment would take for an approximate wavefunction

$$\Psi_{VB} = \{\phi_A(\mathbf{r}_1)\,\phi_B(\mathbf{r}_2) + \phi_A(\mathbf{r}_2)\,\phi_B(\mathbf{r}_1)\}\,\Theta \tag{3.1}$$

where the spin function is

$$\Theta = \{\alpha(s_1)\,\beta(s_2) - \alpha(s_2)\,\beta(s_1)\} \tag{3.2}$$

Goddard's GVB method is a simple extension of this. For a two-electron system, we would write an expression similar to (3.1) above, but allow each 'atomic' orbital ϕ_A, ϕ_B to vary independently until an energy minimum is reached. Obviously, the GVB ϕ_A and ϕ_B will still be expressed as a sum of basis functions.

So far, there appears to be little difference between the GVB and the UHF procedure discussed earlier. As we saw earlier, however, the UHF wavefunction is not a pure spin eigenfunction, so the GVB method neatly resolves the dilemma that doubly occupied orbitals are necessary in order to give the correct spin symmetry whilst singly occupied orbitals are necessary to give correct spectroscopic dissociation products.

3.5.4 Moller–Plesset perturbation theory

We mentioned in Section 1.8 that perturbation theory could be used to study the effect of an applied electric field on a molecule. For perturbation theory to be useful, we need to be able to write the Hamiltonian

$$\hat{H} = \hat{H}_0 + \lambda \hat{H}^{(1)} \tag{3.3}$$

where the zero-order problem can be solved:

$$\hat{H}_0 \Psi_k = E_k \Psi_k \tag{3.4}$$

Moeller–Plesset perturbation theory takes the Hartree–Fock (HF) Hamiltonian as $\hat{H}_0$ and treats as a perturbation the full many-electron Hamiltonian *minus* the HF Hamiltonian. The eigenfunctions of $\hat{H}_0$ are the HF determinant and all singly, doubly, . . . excited states, as in Section 3.5.1. Rayleigh–Schrödinger perturbation theory is then used to obtain the corrected wavefunction and energies through different orders of perturbation theory. If the HF wavefunction is of RHF type we speak of MP2, MP3, . . . calculations (Moller–Plesset perturbation theory to second, third, . . . order), and for a UHF wavefunction we refer to UMP2, UMP3, . . . calculations. All these options are available within GAUSSIAN 86.

From a computational point of view, MP2 calculations are particularly easy because all that is needed is a partial transformation of the two-electron integrals from the atomic to the molecular basis. This is to be compared with CISD calculations, where a full transformation is generally required.

3.6 SIZE CONSISTENCY

Suppose for the sake of argument that we wished to study the dispersion interaction between two helium atoms

2 He → He. . .He

when we would naturally be interested in the energy difference $E(\text{He}\ldots\text{He}) - 2E(\text{He})$. For a straightforward SCF calculation, we find that $E(\text{He}\ldots\text{He}) - 2E(\text{He}) \to 0$ as $R(\text{He}\ldots\text{He}) \to \infty$, but this is *not* the case for certain correlated wavefunctions! Technically, if $E^{(2)}(1)$ and $E^{(3)}(1)$ are second- and third-order energy corrections, then a method is only *size consistent* if these corrections for an ensemble of M isolated identical molecules obey

$$E^{(2)}(M) = ME^{(2)}(1)$$
$$E^{(3)}(M) = ME^{(3)}(1)$$

SCF, MP2 and MP3 calculations are size consistent whilst CISD is not. One solves the problem by defining $E(\text{He})$ to be $\frac{1}{2}E(\text{He}\ldots\text{He})$ with a very large $R(\text{He}\ldots\text{He})$.

CHAPTER 4

Molecular Geometries

4.1 CLOSED-SHELL ELECTRONIC STATES

We wish to predict the bond length of a diatomic molecule, at SCF level. All that is necessary is to repeat the SCF calculation for different values of internuclear separation R and look for the minimum energy. The results of such calculations for a number of first-row closed-shell diatomics are shown in Table 4.1, at the STO/3G and TZVP levels of basis set sophistication.

There are obviously some anomalies, but a more systematic study shows that SCF calculations using quite modest basis sets can give very respectable agreement with experiment and can therefore be used predictively to ordinary chemical accuracy.

The experimental values recorded in the table are themselves subject to various forms of error and may refer to the lowest vibrational level rather than the equilibrium bond distance. Comparisons between theory and experiment are not usually valid below the level of 1 pm/1°.

There are certain systematic deviations between theory and experiment. The STO/3G basis set usually predicts bonds that are too long. Calculations at the minimal basis set level on molecules containing fluorine are usually in rather poor overall agreement with experiment.

The use of a large basis set at the SCF level usually gives bonds which are slightly too short, typically by 1 pm, and this underestimation is more pronounced for bonds involving N, O and F.

4.2 OPEN-SHELL ELECTRONIC STATES

SCF calculations on molecules with open-shell electronic configurations are usually done using the UHF method, despite its drawback of not giving a correct spin eigenfunction. If a low-lying electron state of the same symmetry has a very different geometry to the state under investigation, the calculated geometry may be signifi-

Table 4.1 Geometries of some closed-shell first-row diatomics at the STO/3G and TZVP SCF levels.

	R_e(pm)		
Molecule	STO/3G	TZVP	Expt
BeH^+	132.24	130.92	131.21
BeO	126.90	129.73	133.08
BH	121.25	122.06	123.25
CH^+	118.49	111.18	113.08
CO	114.56	110.47	112.82
F_2	131.46	133.60	143.5
HF	95.53	89.83	91.71
N_2	113.39	106.83	109.4

cantly in error. Table 4.2 shows a selection of results for first-row diatomics. The agreement with experiment is much less impressive, and two general conclusions reached on the basis of more exhaustive studies are that:

(i) for small basis sets, UHF geometries are poorer than RHF;
(ii) for large basis sets, UHF and RHF results are usually similar, provided the expectation value of S^2 $\langle S^2 \rangle$ is reasonable for the UHF calculation.

Thus for example, the very poor UHF results for CO^+ are associated with values of $\langle S^2 \rangle$ equal to 1.463 and 2.166 instead of the correct value, 3/4.

Table 4.2 Geometries of some open-shell first-row diatomics at the STO/3G and TZVP, UHF-SCF levels.

Molecule/state		R_e (pm) STO/3G	TZVP	Expt
B_2	$^3\Sigma_g^-$	153.1	154.3	158.9
BeF	}	129.66	139.87	136.14
BeH	}	130.13	135.19	134.31
BH^+	} $^2\Sigma^+$	120.66	118.56	121.46
BO	}	119.01	118.31	120.49
CN	}	123.5	128.86	117.18
CO^+	}	120.58	137.52	111.50
NH	$^3\Sigma^-$	121.7	102.18	103.8
O_2	$^3\Sigma_g^-$	121.71	116.03	120.74
OH^+	$^3\Sigma^-$	108.35	100.83	102.89

4.3 ELECTRON CORRELATION

The literature contains a large number of geometries calculated using some kind of correlated wavefunction. The simplest treatment possible is to use second-order Moeller–Plesset perturbation theory (MP2). Third-order Moeller–Plesset (MP3) and configuration interaction with double excitations (CID) are also available within GAUSSIAN 86, and it turns out that the latter two approximations yield almost exactly the same results. Table 4.3 contains a representative sample of calculations.

For bonds to hydrogen, adding electron correlation at the MP2 level generally leads to increased bond lengths and better agreement with experiment. Further progress to MP3 leads to a small additional lengthening but little overall improvement in the excellent agreement with experiment.

Bond lengths between non-hydrogen atoms are more strongly altered by the inclusion of electron correlation. We noted above the MP2 calculations for such molecules tended to overestimate the equilibrium geometry. Further progress to MP3 leads to a reduction, giving excellent overall agreement with experiment.

DeFrees *et al.* (1967) reported a statistical analysis of 66 bond lengths which shows that the MP3 and CID techniques give comparable results for equilibrium structures, and that a majority (72%) of calculated parameters lie within the experimental error bars at this level of theory.

Table 4.3 shows a small sample of correlated wavefunction, diatomic molecule geometry predictions for a mixture of closed- and open-shell electronic states.

Table 4.3 Effect of electron correlation on geometry predictions. All calculations using the STO/6-311**G basis set, and all bond lengths pm.

Molecule	HF	MP2	MP3	CID	Expt/pm
H_2	73.8	73.8	74.2	74.6	74.1
LiH	163.6	164.0	164.3	164.9	159.6
BH	122.5	123.4	124.0	124.1	123.2
CH	110.8	110.7	112.0	112.6	112.0
N_2	107.8	113.1	111.6	110.3	109.8
N_2^+	117.7	109.1	108.0	108.5	116.6
O_2^+	116.8	116.3	124.2	124.1	120.8
BeH	134.8	134.6	134.8	135.1	134.3
CH	110.8	110.7	112.0	112.6	112.0
OH^+	101.3	101.1	103.5	103.8	102.8

4.4 LARGER MOLECULES

We wish to predict a molecular geometry for water, which we assume to have C_{2v} symmetry. We therefore need to vary the two parameters R_{OH} and $\theta = H\hat{O}H$ in order to find the energy minimum, and we might decide to do this by:

(a) finding the best R_{OH} for a fixed θ; and then
(b) finding the best θ at this R_{OH}.

This may or may not lead to the minimum. Figure 4.1 shows potential energy surfaces for two triatomic molecules, represented as contour diagrams. The aim of the game is to reach the minimum point. In the case of the left-hand PE surface, the simple strategy outlined above will lead directly to the minimum. In the case of the other PE surface, it will be necessary to repeat steps (a) and (b) several times in order to reach the minimum, simply because of the shape of the PE surface.

What is really needed is the ability to vary both R and θ simultaneously in order to descend the PE surface as quickly as possible. This is a well-known problem in science, engineering and managemenet, and a great deal of effort has gone into the development of efficient numerical algorithms to optimize functions of several variables.

If the PE surface $U(R)$ were a function of a single variable, R, we would naturally ask about the gradient dU/dR. For a function of

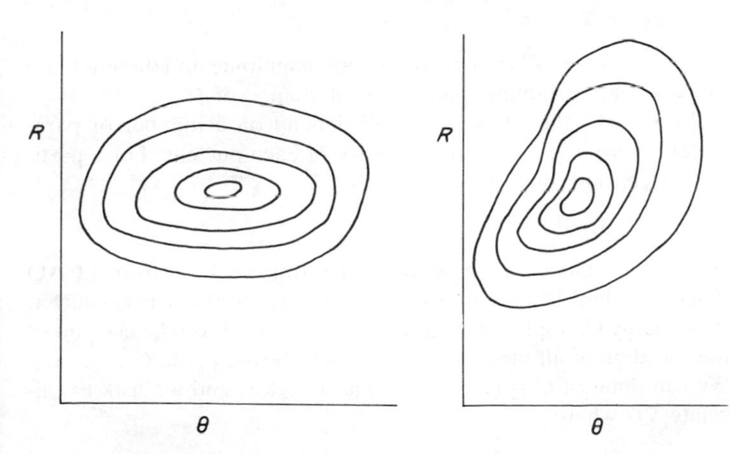

Figure 4.1 Representative potential energy surfaces for two different triatomic molecules

many variables, we need to generalize the idea of gradient. Suppose for the sake of argument that the PE surface depends on three Cartesian coordinates, x, y and z. We need to ask how U changes when we make a differential step from $\mathbf{r} = x\mathbf{i} + y\mathbf{j} + z\mathbf{k}$ to $\mathbf{r} + d\mathbf{r}$ where $\mathrm{d}\mathbf{r} = \mathbf{i}\mathrm{d}x + \mathbf{j}\mathrm{d}y + \mathbf{k}\mathrm{d}z$, $\mathbf{i}, \mathbf{j}, \mathbf{k}$ being unit vectors along the x, y and z Cartesian axes.

We know from elementary calulus that

$$\mathrm{d}U = \frac{\partial U}{\partial x}\mathrm{d}x + \frac{\partial U}{\partial y}\mathrm{d}y + \frac{\partial U}{\partial z}\mathrm{d}z \tag{4.1}$$

and we write this as

$$\mathrm{d}U = \left(\frac{\partial U}{\partial x}\mathbf{i} + \frac{\partial U}{\partial y}\mathbf{j} + \frac{\partial U}{\partial z}\mathbf{k}\right)\cdot \mathrm{d}\mathbf{r} \tag{4.2}$$

where the $\cdot$ signifies a vector dot product.
The vector

$$\frac{\partial U}{\partial x}\mathbf{i} + \frac{\partial U}{\partial y}\mathbf{j} + \frac{\partial U}{\partial z}\mathbf{k}$$

is called the *gradient of* U and is usually written $\boldsymbol{\nabla} U$. The gradient operator $\boldsymbol{\nabla}$ is defined as

$$\boldsymbol{\nabla} \equiv \frac{\partial}{\partial x}\mathbf{i} + \frac{\partial}{\partial y}\mathbf{j} + \frac{\partial}{\partial z}\mathbf{k} \tag{4.3}$$

So

$$\mathrm{d}U = \boldsymbol{\nabla} U \cdot \mathrm{d}\mathbf{r}$$

which shows that $\boldsymbol{\nabla} U$ is a vector whose magnitude and direction are those of the maximum spatial rate of change of U.

In general, the PE surface will depend on a number of parameters, not just the position in space of each nucleus. For a particular electronic state we might write

$$\Psi_0 = \Sigma\, C_k \Psi_k$$

where Ψ_k might be a single Slater determinant built from LCAO functions comprising atomic orbitals centred on the various nuclei. Any energy (U) optimization will therefore need to take account of the variation of all these parameters, which we denote $C_1, \ldots, C_n$. We can think of $\mathbf{C} = (C_1, \ldots, C_n)$ as a vector, and we seek to calculate $\boldsymbol{\nabla} U$ where

$$\boldsymbol{\nabla} U = \frac{\partial U}{\partial C_1}\hat{\mathbf{C}}_1 + \frac{\partial U}{\partial C_2}\hat{\mathbf{C}}_2 + \ldots + \frac{\partial U}{\partial C_n}\hat{\mathbf{C}}_n \tag{4.4}$$

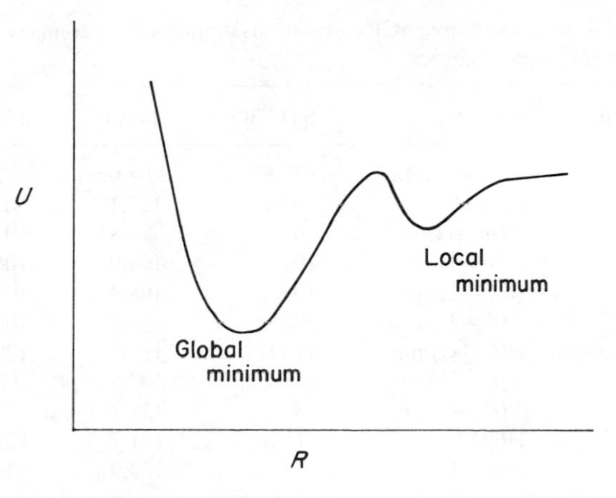

Figure 4.2 A diatomic potential energy surface showing two minima

We refer to techniques involving ∇U as *gradient techniques*. Such techniques have considerable advantages over univariate search techniques, where each variable is optimized independently. In numerical analysis, the choice of gradient technique is determined by whether *analytical* expressions are available for the derivatives, and if the expressions are not available analytically they need to be estimated by finite differences:

$$\frac{\partial U}{\partial C_1} = \frac{U(C_1 + \Delta, C_2, \ldots, C_n) - U(C_1, C_2, \ldots, C_n)}{\Delta} \tag{4.5}$$

A great deal of effort has been expended in evaluating the gradients analytically for SCF, MP2, CID, etc., wavefunctions, but such calculations are inevitably expensive because they require the calculation of all gradient integrals over atomic orbitals.

Potential energy surfaces for polyatomic molecules need not have a single minimum. Figure 4.2 shows a PE surface with two minima. Depending on the starting point in the calculation, either minimum will be reached. We refer to the lowest minimum as the *global* minimum, and to the others as *local* minima.

As we noted for diatomics, basis set dependence does turn out to be important, although minimal basis set calculations can give surprisingly reliable predictions. It is particularly important for gradient calculations to use the smallest possible basis set because of the tremendous cost of the derivative integrals evaluation. Most workers recommend STO/4-31G or STO/4-21G. Calculations of the

Table 4.4 Representative SCF calculations of molecular geometry. Bond lengths pm, angles degrees.

Molecule	Property	STO/3G	TZVP	Experiment
H_2O	R(O—H)/pm	98.94	94.06	95.7
	HOH/°	99.9	107.1	104.5
NH_3	R(N—H)/pm	103.3	99.83	101.2
	HNH/°	104.2	108.9	106.7
CH_4	R(C—H)/pm	108.3	108.3	108.6
C_2H_2	R(C—H)/pm	106.5	105.5	106.1
	R(C—C)/pm	116.8	118.3	120.3
C_2H_4	R(C—H)/pm	108.2	107.5	108.1
	R(C—C)/pm	130.6	131.7	133.4
	HCH/°	115.6	121.7	121.3
CF_4	R(C—F)/pm	136.6	129.9	131.7

energy gradient are twice as costly with the STO/4-31G basis set than the STO/4-21G set. Table 4.4 records representative SCF calculations for a number of small molecules using the STO/3G and TZVP basis sets.

4.5 MOLECULAR MECHANICS

No matter how powerful one's computer, how carefully written one's software or how great one's share of available computer resource, there will always be a problem that is out of reach of *ab initio* techniques. Again, one may be interested more in looking at trends across a series of molecules rather than obtaining spectroscopic accuracy for a geometry prediction. *Ab initio* SCF gradient calculations are expensive calculations. If procedures beyond Hartree–Fock are used, they rapidly become prohibitively expensive calculations.

A traditional alternative is to regard a molecule as a collection of balls and springs, and predict molecular geometries by minimizing a classical potential energy function. In general, we would write the energy U as a function of bond stretching, angle bending, bond torsion, non-bonded and coulombic interactions. The technique is termed 'molecular mechanics', and several packages are available for performing molecular mechanics calculations. *Hydrocarbon* force fields are well parameterized, but the parameterization for other elements is less reliable.

An example of a currently-used alkene force field is given by White and Bovill (1977).

$$
\begin{aligned}
U = {} & \tfrac{1}{2}\Sigma k_l(l - l_0)^2 - \tfrac{1}{2}\Sigma k_\theta(\Delta\theta^2 - k'_\theta\Delta\theta^3) \\
& + \tfrac{1}{2}\Sigma k_\omega(1 + S\cos n\omega) \\
& + \Sigma \epsilon_0\left\{-\frac{2}{\alpha^6} + \exp(12(1 - \alpha))\right\} \\
& + \tfrac{1}{2}\Sigma k_\chi(180 - \chi)^2
\end{aligned}
\tag{4.6}
$$

where $\alpha = r/(r_1^* + r_2^*)$ with r^* a van der Waals radius and χ an improper torsion angle which is used to account for out-of-plane bonding at trigonal atoms. The first term represents bond stretching, the second angle bending, the third bond torsion, the fourth non-bonded interactions.

Molecular mechanics can be applied to single molecules and to aggregates (e.g. crystals). Particular attention has to be paid to the choice of optimization technique, as one may be optimizing a function of (say) 100 variables. Most of the traditional techniques suffer from convergence problems.

Figure 4.3 shows output from a typical molecular mechanics calculation of the preferred conformers of the C_6–C_{10} cycloalkanes.

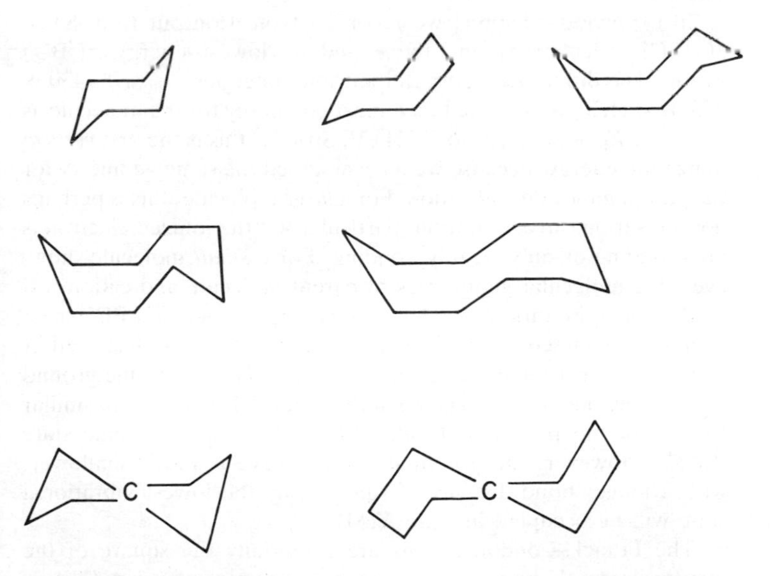

Figure 4.3 Representative molecular mechanics conformational analysis calculations

CHAPTER 5

Energies

5.1 KOOPMANS' THEOREM

The interaction of a molecule M with photons of energy less than about 8 eV leads to excitation of the valence electrons, and the molecular absorption spectrum can be measured with a visible or ultraviolet spectrometer. Increasing the photon energy leads to photoemission of electrons, and the different ionic states are usually studied by photoelectron spectroscopy (PES). In ultraviolet PES (UVPES) we use typically HeI radiation of energy 21.2 eV, whilst in X-ray PES we use X-rays having typically energies of >1 keV. UVPES ionizes valence electrons, whilst XPES ionizes inner-shell electrons.

In a previous chapter, we recorded typical output from STO/3G SCF calculations on ethene and its lowest energy $\pi(^2B_{3u})$ cation. The difference between their total energies $(-76.785\,430 + 77.069\,401)\,E_h$ gives us the first ionization energy for the molecule as $0.283\,971\ E_h$, equivalent to 7.724 eV. Strictly, this is the first *vertical* ionization energy, because we have assumed the same geometry for the parent molecule and cation. For a *large* molecule, this is perhaps not too serious an assumption, particularly if the ionized electron is non-bonding or only weakly bonding. For a *small* molecule, however, the molecular geometries of parent molecule and cation will tend to be quite different. The physical appearance of a PES spectrum is determined by the Franck–Condon factors, as illustrated in Figure 5.1 for a typical diatomic molecule. The electronic ground state of the ion $X(M^+)$ is such that the potential curve is very similar to that of the parent molecule, $X(M)$. For the electronic state $\tilde{A}(M^+)$, however, the potential energy curve is much shallower, with a longer bond distance. Usually, only the lowest vibrational state will be occupied in state $X(M)$.

The Franck–Condon factors are essentially the square of the overlap integrals between the upper and lower *vibrational* wavefunctions, and since ionization is essentially an instantaneous

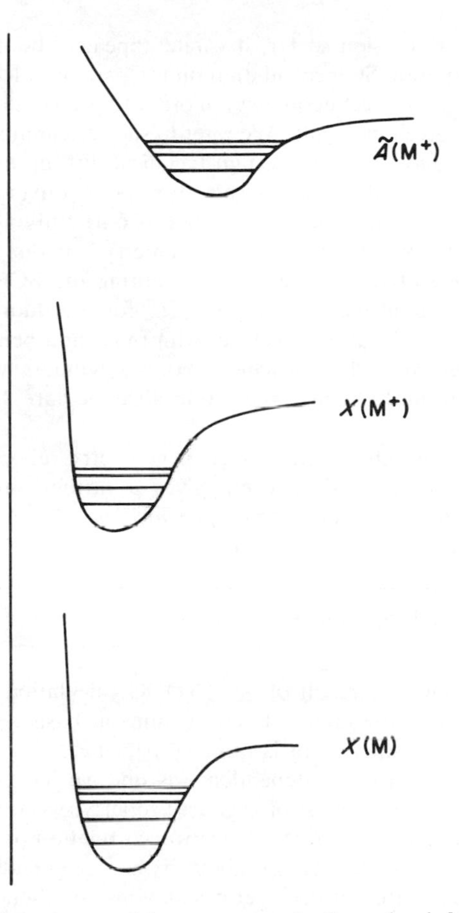

Figure 5.1 Diatomic potential energy curves for the molecule M and two of its ions

process, the appearance of the two vibrational envelopes for the two ionization processes will be quite different.

Photoelectron spectroscopy often gives both the adiabatic and vertical ionization energies. An adiabatic ionization energy is defined as an energy difference with both neutral molecule and ion in their lowest vibrational and rotational states, whilst a vertical ionization energy is an energy difference when the molecule and the ion have the same geometries. Obviously the spectra of polyatomic species are more complicated than diatomics because many different vibrational modes need to be considered.

From the discussion so far, it would appear to be necessary to perform a separate SCF calculation on the parent molecule and on each ion (at the correct geometry) in order to predict the ionization energies of a given molecule. We refer to such calculations as ΔSCF calculations, and there are certain technical difficulties associated with them. Firstly, they usually show very poor convergence in the SCF cycles. Secondly, the SCF method is only truly applicable to the lowest energy state of any given symmetry, and one has to force the occupancy of the desired orbitals during the SCF iterations. Thirdly, such a calculation on (say) CF_4 for the fluorine 1*s* ionization gives a 'delocalized' picture, with ionization being from the four fluorines. In order to achieve good agreement with experiment, it turns out to be necessary to localize the core 'hole' on *one* atom.

Most workers ignore the effect of geometry relaxation in the cation, and Koopmans' theorem gives a further simplification. Koopmans' theorem states that, *provided the orbitals of parent molecule and cation are identical*,

$$\boxed{\text{Ionization energy} = -\text{ orbital energy}}$$

Table 5.1 shows the result of our STO/3G calculation for ethene, compared with experiment. Even at minimal basis set level, the agreement with experiment is quite convincing.

The idea of basis set dependence is one we have touched on earlier; for the calculation of a given property, we should always ask about the probable effect of electron correlation (if we are only doing an SCF calculation) and about basis set dependence. Table 5.2 shows four different basis set calculations on ethene. The basis set dependence is not severe. As a rule of thumb, however, it is

Table 5.1 Koopmans' theorem prediction of ionization energy (eV) at the STO/3G level, compared to experiment.

Ionic state	Koopmans'	Expt
$^2B_{3u}$	9.41	10.51
$^2B_{3g}$	11.62	12.46
^2A_g	15.45	14.46
$^2B_{2u}$	15.59	15.78
$^2B_{1u}$	18.53	18.87

Table 5.2 Orbital energies of ethene using different basis sets.

Orbital	STO/3G	STO/6G	SV/4-31G	TZVP
1	−11.018	−11.171	−11.208	−11.232
2	−11.017	−11.170	−11.207	−11.231
3	−0.976	−0.978	−1.028	−1.032
4	−0.748	−0.751	−0.789	−0.797
5	−0.609	−0.612	−0.647	−0.652
6	−0.523	−0.525	−0.573	−0.580
7	−0.473	−0.476	−0.509	−0.519
8	−0.319	−0.322	−0.366	−0.371

necessary for the (experimental) ionization energies to be separated by about 2 eV to be able to place any reliance on Koopmans' theorem assignments.

Koopmans' theorem fails dramatically for some very simple molecules; the textbook example is N_2, where the experimental ordering of ionizations is quite different from that calculated from Koopmans' theorem.

The effect of electron correlation has been widely investigated, but the fact is that Koopmans' theorem is today widely used for systems of chemical interest, because of the lucky cancellation of relaxation and correlation effects.

5.2 POTENTIAL ENERGY SURFACES

As we noted earlier, the concept of a potential energy surface owes its existence to the Born–Oppenheimer theorem, where the (electronic and nuclear) wavefunction is factored into a product of electronic and nuclear parts. The nuclei are fixed in space for the purpose of calculating the *electronic* wavefunction, and the nuclei then move in the potential generated by the electrons. Potential energy (PE) surfaces play a dominant role in spectroscopy, in reaction kinetics, in computer simulation and in scattering, and PE surfaces are therefore of interest to both theoreticians and experimentalists.

Spectroscopists usually fit experimental data to some semi-empirical function, the two most widely used potentials for diatomics being the Morse function

$$V(R) = D_e\{\exp(-2\alpha(R - R_e)) - 2\exp(-\alpha(R - R_e))\} \quad (5.1)$$

and the Lennard–Jones 12-6 potential

$$V(R) = 4\varepsilon\{(\sigma/R)^{12} - (\sigma/R)^{6}\} \quad (5.2)$$

For a polyatomic molecule, potential energy surfaces are much more complex, and we return to this problem shortly.

Let us consider for the moment a prototype reaction:

$$H_2 \rightarrow 2H$$

Experimentally, the dissociation process requiring least energy is dissociation into two ^2S ground-state hydrogen atoms, and we know from spectroscopic data that the dissociation energy of H_2 is 4.72 eV ($0.173E_h$) at $R_e = 74.16$ pm. Table 5.3 shows the result of an SCF calculation of the potential energy curve, using a large basis set. It is quite clear that there is a problem: with this particular basis set, $E(H^-) = -0.467\,392E_h$ and $E(H) = -0.499\,809\ E_h$, and it can be seen that the dissociation products are incorrectly predicted by the SCF calculation. The SCF calculation refers actually to

$$H_2 \rightarrow \tfrac{1}{2}(2H(^2S) + H^+ + H^-)$$

and the predicted dissociation energy is also very poor. Unfortunately, this is typical behaviour where the 'products' in

$$AB \rightarrow A + B$$

have unpaired electrons. Some treatment of electron correlation is usually essential.

A large number of PE surfaces involving two, three and four atomic combinations have been compiled, despite the fact that such

Table 5.3 Calculated potential energy for H_2 at the TZVP SCF level.

R/a_0	E/E_h
1.0	−1.083 192
1.2	−1.123 779
1.4	−1.132 595
1.6	−1.124 283
1.8	−1.109 740
2.0	−1.090 177
3.0	−0.986 437
4.0	−0.764 081
5.0	−0.856 297
6.0	−0.821 259
10.0	−0.764 081

surfaces have to be calculated point by point. Clearly, the wavefunction has to be suitable for describing the asymptotic regions of the surface such as dissociation to atom + molecule or to three atoms (in the case of a triatomic). For most PE surfaces, a treatment of electron correlation is essential. The choice of basis set is an independent matter: electron correlation is needed in order that the wavefunction shows the correct dissociation products.

A well-studied example is H_3: the reaction

$$H_2 + H \rightarrow H + H_2$$

is of great importance in chemistry as it is the simplest exchange reaction involving neutral species. Liu and Siegbahn reported CI calculations with a large basis set, and confirmed that the minimum energy path was the collinear one. They calculated energies for 156 geometries on the three-dimensional potential surface, and their results are believed to be accurate to 0.5 kJ mol^{-1}, which is the accuracy needed for most chemical applications.

Potential energy surfaces involving four or more atoms require six or more internal coordinates to describe the relative positions of the nuclei, and obviously the cost associated with generating such surfaces is prohibitive. It is usual to try and predict which degrees of freedom are relevant to a particular chemical reaction using chemical intution, and this is discussed in some detail in Bader and Gangi's review.

5.3 CORRELATED WAVEFUNCTION CALCULATIONS

In order to understand the principles involved, consider the reaction

$$CH_2(^3B_1) \rightarrow C(^3P) + H_2(^1\Sigma_g^+)$$

The electronic ground-state configuration of CH_2 is $1a^22a^21b_2^2$ $3a_1^11b_1^1$. On dissociation, we need to determine which CH_2 molecular orbitals correlate with the orbitals of the products, and a simple consideration of symmetry properties shows us that C 1s correlates with $1a_1$, C 2s with $2a_1$, the hydrogen $1\sigma_g$ orbital correlates with $3a_1$, C $2p_z$ with $4a_1$ and C $2p_x$ with $1b_1$. Hence we need to include at least *two* reference states to describe the dissociation of CH_2 along this pathway; one for each of

$1a_1^2 2a_1^2 1b_2^2 3a_1^1 1b_1^1$ and
$1a_1^2 2a_1^2 3b_1^2 4a_1^1 1b_1^1$.

5.4 SEMI-EMPIRICAL PE SURFACES

A *semi-empirical* calculation is one which relies partly on experimental data for calibration, and a simple example of such a calculation is the PE surface for H_3 constructed by London, Eyring and Polanyi (LEP) (Eyring and Polanyi, 1931). In the event of negligible overlap between three atomic 1*s* orbitals, LEP's simple extension of the valence bond (VB) treatment of H_2 gives two states:

$$E_{\pm} = J_{ab} + J_{ac} + J_{bc} \pm [(K_{ab} - K_{ac})^2 + (K_{ab} - K_{bc})^2 + (K_{bc} - K_{ac})^2] \quad (5.3)$$

where the J and K are the 'Coulomb' and 'exchange' integrals of VB theory. In those regions of the PE surface where overlap is important, the energy expressions a very much more complicated, and Sato (1955) proposed that the effect of overlap could be simulated by simply multiplying $E_{\pm}$ by $(1 \pm k)^{-1}$, where the constant k is determined 'experimentally'.

There has been a resurgence of interest in the last decade in the use of *ab initio* techniques to help in the construction of semi-empirical potential surfaces. The general objective is to use a small number of *ab initio* calculations to parameterize the interaction potential, which we now discuss.

For two atoms A, B we define the incremental pair potential $\Delta U^{(2)}(\text{A, B})$ as

$$\Delta U^{(2)}(\text{A, B}) = U(\text{A, B}) - U(\text{A}) - U(\text{B}) \quad (5.4)$$

(By $U(\text{A})$ we mean the total energy of A in the absence of B).

$\Delta U^{(2)}(\text{A, B})$ is obviously just the same as the potential energy of Figure 5.1. For a triatomic molecule ABC, we will focus attention on the interaction potential

$$\Delta U(\text{A, B, C}) = U(\text{A, B, C}) - U(\text{A}) - U(\text{B}) - U(\text{C}) \quad (5.5)$$

and we are interested in how this interaction potential is related to $\Delta U^{(2)}$ for each possible diatom AB, AC and BC. We write

$$\Delta U(\text{A, B, C}) = \Delta U^{(3)}(\text{A, B, C}) + \Delta U^{(2)}(\text{A, B}) + \Delta U^{(2)}(\text{A, C}) + \Delta U^{(2)}(\text{B, C}) \quad (5.6)$$

where the three-body term $\Delta U^{(3)}(\text{A, B, C})$ owes its existence to the polarization of each atom by the others. If the atoms were point charges, $\Delta U^{(3)}(\text{A, B, C})$ would equal zero.

For an n-atom molecule we look for a decomposition of the interaction potential into two-body, three-body, . . . interactions

$$\Delta U(\text{A, B}, \ldots, \text{N}) = \Delta U^{(n)}(\text{A, B}, \ldots, \text{N}) + \Sigma \Delta U^{(n-1)} + \ldots + \Sigma \Delta U^{(2)} \quad (5.7)$$

and the aim of modern semi-empirical potential surface calculations is to give a very simple functional representation to $\Delta U^{(n)}$.

5.5 WHEN ARE SCF CALCULATIONS ADEQUATE?

If one is interested in the region of a PE surface around an equilibrium position, SCF calculations offer remarkably good predictions for a wide range of molecular properties. We have just seen that SCF calculations are not *in general* adequate for the construction of PE surfaces, because of the incorrect dissocation products.

There are, however, certain types of system which constitute exceptions to this rule. The first important example is one in which the electronic structures of the dissociation products correspond to closed-shell systems, e.g.

$$KF \rightarrow K^+ + F^-$$

and Table 5.4 shows typical SCF calculations for a selection of alkali halides. The SCF wavefunction gives the correct dissocation products.

A more complicated example is

$$CN^-(^1\Sigma^+) + CH_3F(^1A) \rightarrow CH_3CN(^1A) + F^-(^1S)$$

Table 5.4 Dissociation energies for a selection of alkali halides at the SCF level using large polarized Dunning basis sets. In all cases, the dissocation products are *ions*.

Molecule (X—Y)	R_e(X—Y) (pm)		D_e(kJ mol^{-1})	
	Calc.	Expt.	Calc.	Expt.
LiF	156.9	156.4	771.3	770.3
LiCl	202.4	202.1	628.3	641.4
LiBr	220.0	217.0	595.9	618.9
NaF	192.7	192.6	643.5	643.9
NaCl	242.1	236.1	537.9	554.8
NaBr	253.9	250.2	519.2	534.3
KF	224.9	217.1	570.0	582.4
KCl	266.7	266.7	471.8	493.7
KBr	293.5	282.1	453.0	475.3

and a number of studies have been reported for such systems. In each case, the reactive complex has C_3 symmetry. In such systems, the number of electron pairs remains constant, and so Bader and Gangi (1976) note that '. . .one anticipates that major changes in the correlation energy will occur only when the number of electron pairs predicted by the orbital model of the system changes during the course of the reaction'. Snyder and Basch (1969) claim to have confirmed this prediction for an extensive series of reactions, and these authors pay particular attention to comparison between experimental and SCF standard enthalpy change for the reaction under study. In comparing theory with experiment, it is necessary to convert internal energies to enthalpies, and to allow for the translational, rotational and vibrational degrees of freedom. These effects are usually allowed for very roughly by assuming ideal gas behaviour, when

$$\Delta H = \Delta U + RT\Delta n \tag{5.8}$$

with Δn the change in the number of gaseous moles, the rotational and translational internal energies having their classical values. The vibrational internal energy is usually taken as $h\omega_i$ for each normal mode; the frequencies can be either *experimental* or based on the SCF calculation.

A word of caution is appropriate. We take two specific examples from Snyder and Basch's work (1969):

$$H_2 + F_2 \rightarrow 2HF$$
$$HCN + H_2O \rightarrow CO + NH_3$$

Table 5.5 shows the SCF energies and spectroscopic quantities for all molecules, all SCF calculations referring to the same basis set and the experimental geometry. The predicted enthalpy changes work out as -1875 and $+1613$ kJ mol^{-1} respectively, for comparison with the experimental values of -542.2 ± 16.7 and -45.44 ± 3.37 kJ mol^{-1} respectively. On the basis of near Hartree–Fock energies, the author's claim much better agreement with experiment.

Snyder and Basch's calculations (1969) established that enthalpies of *hydrogenation* are well represented at SCF level. They thus report calculations for reactions such as

$$H_2 + C_2H_6 \rightarrow 2CH_4$$
$$2H_2 + C_2H_4 \rightarrow 2CH_4$$

Hehre, Ditchfield, Radom and Pople (1970) found that it is more reliable to describe the hydrogenation in two steps:

Table 5.5 Energies for the reactions $HCN + H_2O \rightarrow CO + NH_3$ and $F_2 + H_2 \rightarrow 2HF$ all calculated with experimental geometry at the SCF TZVP level. Vibrational contributions are experimental values.

	E_{SCF}	E_{vib}	E_{rot}	E_{trans}	E_{tot}/E_h
F_2	−198.2667	0.0020	0.0009	0.0014	−198.2624
H_2	−1.1326	0.0099	0.0009	0.0014	−1.1204
HF	−100.0601	0.0093	0.0009	0.0014	−100.0485
					−0.7142
HCN	−92.9043	0.0155	0.0009	0.0014	−92.8865
H_2O	−76.0553	0.02036	0.0014	0.0014	−76.0319
CO	−112.1306	0.0049	0.0009	0.0014	−112.1234
NH_3	−56.2163	0.0327	0.0014	0.0014	−56.1808
					0.6142

(i) the reaction in which all bonds between heavy atoms are broken to give the simplest possible molecule with each kind of bond; and then
(ii) the full hydrogenation of all products.

Thus using $CH_3CH{=}C{=}O$ as an example, step (i) is

$$CH_3-CH{=}C{=}O + 2CH_4 \rightarrow C_2H_6 + C_2H_4 + H_2CO$$

where the left-hand side methane has been added to achieve stoichiometric balance. Step (ii) is

$$C_2H_6 + H_2 \rightarrow 2CH_4$$
$$C_2H_4 + 2H_2 \rightarrow 2CH_4$$
$$H_2CO + 2H_2 \rightarrow CH_4 + H_20$$

and adding steps (i) and (ii) gives

$$CH_3-CH{=}C{=}O + 5H_2 \rightarrow 3CH_4 + H_2O$$

5.5.1 Isodesmic reactions

An *isodesmic* reaction is one where the number of bonds of each type is the same in reactants and products. An example is $CH_2{=}CH-CH_2OH + CH_2{=}O \rightarrow CH_2{=}CHOH + CH_3CH{=}O$ because both reactants and products contain 7 @ CH, 1 @ C=C,

1 @ C—C, 1 @ C—O and 1 @ O—H bond, and again a number of these isodesmic reactions have been studied. SCF calculations are generally thought to be adequate for such reactions.

5.6 BARRIERS TO INTERNAL ROTATION AND INVERSION

In the mid-nineteenth century it was believed that free rotation could occur around any single bond, but rotation around a multiple bond was restricted. As time passed, the the concept of free rotation around a single bond became suspect, and there is to be found in the current chemical literature an impressive compilation of experimentally determined potential barriers hindering rotation about single bonds, usually carbon–carbon bonds.

Most experimental values have been determined from microwave spectroscopy. Perhaps the best known value is 12.25 ± 0.11 kJ mol^{-1} for ethane, given by Weiss and Leroi (1968). They fitted a torsional potential

$$V - V_0 = V_3(1 - \cos 3\theta) + V_6(1 - \cos 6\theta) + \ldots \quad (5.9)$$

and concluded that only the V_3 term was necessary.

For molecules such as H_2O_2 and N_2H_4, etc., where the symmetry is lower, it is necessary to take

$$V - V_0 = V_1(1 - \cos \phi) + V_2(1 - \cos 2\phi) + V_3(1 - \cos 3\phi) \quad (5.10)$$

and for asymmetric molecules it is necessary to add additional terms in order to reflect the lack of symmetry about $\phi = 180°$.

Returning to ethane, an SCF calculation of the barrier is at first sight particularly easy. Calculate the energies of the staggered and eclipsed conformations and subtract:

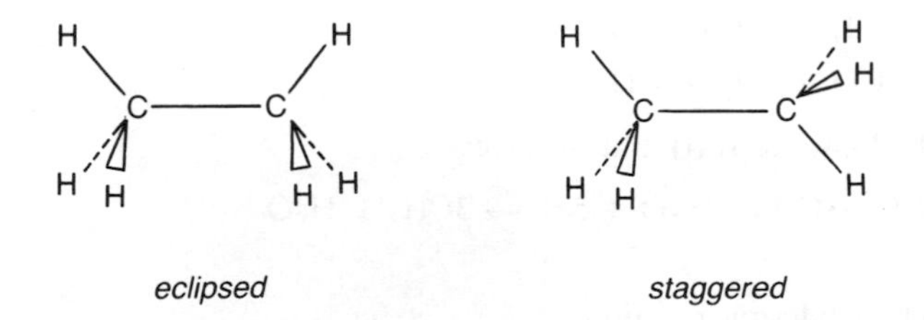

eclipsed *staggered*

Using standard bond lengths and angles (R_{CC} = 154.3 pm, R_{CH} = 110.2 pm, tetrahedral angles) gives values for three choices of basis set, STO/3G, STO/4-31G and TZVP of 16.40, 16.39 and 16.97 kJ mol^{-1}.

Workers in this field usually emphasize the importance of geometry optimization. One usually finds that the eclipsed structure is slightly more 'open' than the staggered one, and this *reduces* the predicted barrier by a few per cent. The corresponding values and geometries are shown in Table 5.6. Although at first sight all basis sets perform equally well for ethane, the use of an extended basis set is recommended.

Electron correlation is thought to make a negligible contribution to rotational barriers about single bonds, but that is not thought to be the case for rotation about bonds such as the N—N bond in N_2O_4 nor the central C—C bond in 1,3-butadiene.

The ammonia molecule is pyramidal, with an $H\hat{N}H$ bond angle of 106°. One mode of vibration can be visualized as where the nitrogen atom moves to and fro through the plane of the hydrogen atoms. This process is known as *inversion*, and the energy difference between the planar and bent structures defines the *barrier to inversion*.

The inversion barrier in ammonia is known to be 24.3 kJ mol^{-1} and on the basis of our discussion of barriers to internal rotation one might anticipate that an extended basis set SCF calculation with geometry optimization at the bent and planar conformations would give good agreement with experiment. Typical SCF calculations are shown in Table 5.7. Even at TZVP level the agreement with experiment is poor and many authors have claimed that this poor agreement was due to the effect of electron correlation. Barriers to inversion turn out to need much larger basis sets, and the inclusion of a double set of polarization functions leads to excellent agreement with experiment.

Table 5.6 The barrier to internal rotation V_3 in ethane. θ is the out-of-plane CH angle.

	R(C—C) (pm)	R(C—H) (pm)	θ (°)	V_3 (kJ mol^{-1})
STO/3G				
Staggered	150.0	109.1	69.9	13.06
Eclipsed	150.6	109.1	69.5	
STO/4-31G				
Staggered	147.9	109.0	69.7	13.01
Eclipsed	148.6	109.0	69.3	
TZVP				
Staggered	148.7	109.1	69.6	12.49
Eclipsed	149.5	109.0	69.1	

Table 5.7 Calculated geometries and barriers to inversion of ammonia.

Basis set		Bent	Planar	Barrier ($kJ\ mol^{-1}$)
	R(N—H) (pm)	103.3	100.5	
STO/3G	$H\hat{N}H$	104.2°	120°	46.6
	E/E_h	−55.455 420	−55.437 665	
	R(N—H) (pm)	99.1	98.6	
STO/4-31G	$H\hat{N}H$	115.7°	120°	1.82
	E/E_h	−56.106 692	−56.105 997	
	R(N—H) (pm)	99.2	98.7	
TZV	$H\hat{N}H$	115.6°	120.0	1.89
	E/E_h	−56.187 818	−56.187 097	

5.7 HYDROGEN BONDED COMPLEXES

When a covalently bound hydrogen atom forms a second bond to another atom or region of high electron density (e.g. the C=C bond in ethene), the second bond is referred to as a hydrogen bond. Hydrogen bonds are for the most part weaker than normal covalent bonds, and can be classified according to the enthalpy change for the 'reaction'

$$A—H + B \rightarrow A—H\ldots B$$

with 10–50 $kJ\ mol^{-1}$ for 'normal' hydrogen bonding.

Credit for the discovery of hydrogen bonding is usually given to Latimer and Rodebush for their 1920 paper, but the concept is undoubtedly much older. We have obviously come a long way since Latimer and Rodebush's formulation of the water dimer as

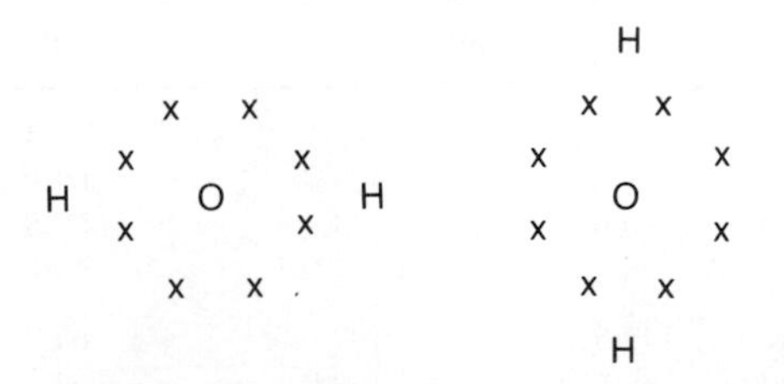

but surprisingly most of the progress has been made in the last decade. Very many experimental studies of hydrogen bonding are in the literature, but the vast majority of these older studies gave very little *molecular* information. New techniques in spectroscopy (such as pulsed nozzle Fourier transform microwave spectroscopy and molecular beam electric resonance) have recently begun to give molecular information such as geometries, electric dipole moments, etc. In the case of HCN. . . HF, it has also proved possible to deduce a spectroscopic hydrogen bond D_0 and also to estimate D_e.

5.7.1 Basis set dependence

The crucial question is as usual that of the choice of atomic orbital basis set. Table 5.8 shows this sensitivity of geometry and hydrogen bond energy to differences in basis set. For illustration, three different degrees of sophistication were chosen, STO/3G, STO/4-31G, and TZVP.

It is a normal experience that the 'monomer' geometries of A—H and B are usually almost exactly retained on formation of the 'dimer' A—H. . . B, usually with a very minor lengthening of the A—H bond. Full geometry optimizations are not necessary for a study of such complexes. Minimal basis set calculations in general are not particularly reliable for either geometry or more seriously for the hydrogen bond energy.

Slightly extended basis sets of the STO/4-31G type generally overestimate the hydrogen bond energy, and this can be rationalized on the argument that the degree of variational freedom within the complex is proportionately higher than in the two isolated monomer units, because each monomer unit can use the basis functions centred on the other unit to improve its own electron density on dimer formation.

It is particularly gratifying to see such an excellent accord between the experimental hydrogen bond energy and that calculated using the largest basis set. The accord between the calculated and experimental *geometries* is respectable for all basis sets. The experimental value is an R_S and so is not directly comparable with the calculated values.

Table 5.9 shows a selection of linear hydrogen-bonded complexes. Not all have been observed experimentally, but experi-

Table 5.8 SCF-MO calculations on HCN. . . HF.

Basis set	R(N. . . H) (pm)	R(H—F) (pm)	D (kJ mol^{-1})
STO/3G	207.9	95.5	15.3
STO/4-31G	192.8	92.2	38.8
TZVP	197.2	90.5	27.5
Experiment	R(N. . . F) = 279.2 pm		$D_e = 26.1 \pm 1.6$

Table 5.9 Calculated and experimental hydrogen bond lengths and energies; all SCF calculations at TZVP level.

Complex		R(C...X) (pm)		D_e(C...H) (kJ mol^{-1})	
ABC...HX		Calc.	Expt.	Calc.	(Expt.)
HCN	HF	287.7	279.2 (R_o)	27.5	(26.1 ± 1.6)
	HCl	352.8	340.5	15.7	
	HBr	374.6	361	12.8	
SCO	HF	274.3	296	9.2	
	HCl	360.3	—	5.5	
OC	HF	303.2	304.7	9.5	
	HCl	360.3	369.4	4.1	
CO	HF	323.6	—	7.9	
	HCl	398.4	—	3.9	
N_2	HF	316.1	308.2	6.7	
	HCl	379.3	—	2.7	
	HBr	422.8	—	2.2	

mental geometries and bond energies are all included where these are available. Only one D has been deduced to date, and in view of our earlier comments it is probable that the SCF results are reliable predictors for all but the weak complexes (those with $D_e < 10$ kJ mol^{-1}). Thus for example, the effect of electron correlation upon potential well depth was investigated at the MP2, MP3 and CID levels of theory using the standard GAUSSIAN 80 6-311G** and 6-31G* basis sets. With the latter basis set D is calculated as 6.18, 11.26, 9.80 and 7.99 kJ mol^{-1} for SCF, MP2, MP3 and CID. Taking the CID result, this means that electron correlation accounts for almost one-third of D.

The sign of the CO electric dipole moment proved an elusive piece of experimental data until Rosenblum, Nethercot and Townes (1958) were able to demonstrate from observation of the effect of isotope substitution upon the *rotational magnetic* moment that it had the sense C^-O^+. Early minimal basis set calculations gave the correct sign (for all the wrong reasons) but any calculation at the SCF level using a respectable basis set gives C^+O^-. The interesting thing about the complexes of CO with hydrogen halides is that the correct atomic order is predicted OC...HX even though the SCF calculation give the polarity of CO as C^+O^-. We do predict stable

CO...HX complexes, but these have not been observed experimentally. The interaction is *not* a simple dipole–dipole one. Obviously there will be some effect of electron correlation on this result, and this effect will presumably be to stabilize the OC...HX complexes relative to CO...HX.

5.8 GHOST ORBITALS

We discussed earlier a calculation of the hydrogen bond energy

$$\text{HCN} + \text{HF} \rightarrow \text{HCN...HF}$$

using the 'supermolecule' approach; the hydrogen bond energy is given directly as

$$\Delta U = U(\text{HCN...HF}) - U(\text{HCN}) - U(\text{HF})$$

and it is invariably the very small difference of two very large numbers. A problem which often arises, particularly when using small basis sets, is that the dimer energy is lower than it should 'really' be, because the HCN fragment in HCN...HF will try to use the orbitals of HF to improve its own electron density and vice versa. Boys and Bernardi (1970) suggested that this problem could be resolved by calculating ΔU a different way. ΔU is given by

$$\Delta U = U(\text{HCN...HF}) - \tilde{U}(\text{HCN}) - \tilde{U}(\text{HF})$$

where the energy $\tilde{U}(\text{HCN})$ refers to an SCF calculations on HCN *in the presence of the HF atomic orbitals*. Such orbitals are usually referred to as ghost orbitals, and this counterpoise correction is generally thought desirable in the case of weak interactions and small basis sets. For HCN...HF at the STO/3G and STO/4-31G levels, the calculated quantities are given in Table 5.10.

Table 5.10 Result of ghost orbital calculations of the hydrogen bond energy in HCN...HF.

	STO/3G	STO/4 – 31G
$U(\text{HF})/E_h$	−98.572 846	−99.887 287
$\tilde{U}(\text{HF})$	−98.572 996	−99.888 647
$U(\text{HCN})$	−91.672 926	−92.731 433
$\tilde{U}(\text{HCN})$	−91.675 433	−92.732 419
$U(\text{HCN...HF})$	−190.251 207	−192.633 468
D_e (kJ mol^{-1})	14.27	38.72
D_e(ghost)	8.33	32.56

CHAPTER 6

Force Fields

A number of chemical and spectroscopic processes involve the relative motion of atomic nuclei. As we saw in a previous chapter, the complete characterization of a PE surface is a very complex task. Often, however, we are interested in the motion of the nuclei in the vicinity of a single point on the PE surface (the global minimum, say), and the PE can usually be expressed as a power series expansion about that point.

If $q_1, \ldots, q_n$ are internal coordinates suitable for describing the nuclear motion (they might be normal coordinates from infrared studies, for example), we can write

$$U = U_0 - \Sigma \phi_k q_k + \tfrac{1}{2}\Sigma\Sigma F_{ij} q_i q_j + \tfrac{1}{6}\Sigma\Sigma\Sigma F_{ijk} q_i q_j q_k + \ldots \tag{6.1}$$

where the forces ϕ_i, quadratic force constants F_{ij}, cubic force constants $F_{ijk}, \ldots$ have to be calculated at the reference point

$$\phi_i = -\left(\frac{\partial U}{\partial q_i}\right)_0 \qquad F_{ij} = \left(\frac{\partial^2 U}{\partial q_i \partial q_j}\right)_0 \ \ldots \tag{6.2}$$

6.1 DIATOMIC MOLECULES

Diatomic molecules are essentially a law unto themselves. Spectroscopic term values are usually written

$$\begin{aligned} T_{v,J} = {} & \omega_e(v + 1/2) - \omega_e x_e(v + 1/2)^2 + \ldots \\ & + B_v J(J + 1) - D_v J^2(J + 1)^2 \ldots \end{aligned} \tag{6.3}$$

where $B_v = B_e - \alpha_e(v + 1/2)$ and $\omega_e = \dfrac{1}{2\pi c}\sqrt{\dfrac{k}{\mu}}$ in the usual notation.

The traditional way of computing force constants for diatomic molecules is to calculate $U(R)$ for a number of different R's, and then to differentiate $U(R)$ *twice* by numerical methods. Numerical differentiation is at best fraught with pitfalls and is in any case a numerically unstable operation.

The difficulties are particularly serious in this case because the first differentiation needs small differences in very large energy values.

A second alternative is to solve numerically the vibrational Schrödinger equation

$$\left(-\frac{1}{2\mu}\nabla^2 + U(R)\right)\Psi_{\mathrm{vib}} = E_{\mathrm{vib}}\Psi_{\mathrm{vib}} \tag{6.4}$$

where μ is the reduced nuclear mass. If *rotational* energy levels are required, it is necessary to include the term $J(J + 1)/2R^2$.

Finally, a Dunham analysis can be used, in which the PE is given as a power series in the variable $\varrho = (R - R_e)/R_e$.

$$U(\varrho) = A_0[1 + \Sigma a_i\varrho^i] \tag{6.5}$$

and A_0, a_i determined by least-squares fitting. Dunham (1932) made a careful study of the fine interactions between diatomic vibrations and rotations. If we express diatomic term values

$$T_{v,J} = \Sigma Y_{l,j}(v + 1/2)^l J^j(J + 1)^j \tag{6.6}$$

where the first Y subscript refers to the *vibrational* level and the second to the *rotational*, Dunham showed that (for example)

$$Y_{1,0} = \omega_e\{1 + B_e^2/(4\omega_e^2)(25a_4 - 95a_1a_3/2 - 67a_2^2/4 + 459a_1^2a_2/8 - 1155a_1^4/64)\}$$

Luckily, the 'correction' terms in curly brackets are usually only different from 1 by much less than 1 part in 1000, and we find that

$$Y_{1,0} = \omega_e \quad Y_{2,0} = x_e\omega_e \quad Y_{0,1} = B_e \quad Y_{0,2} = D_e \quad Y_{1,1} = \alpha_e \ldots$$

It is well to remember that spectroscopists measure differences between term values. These wavenumbers then have to be fitted to an expression similar to equation (6.3), and the spectroscopic 'constants' ω_e, $\omega_e x_e$, etc., depend critically on where the experimentalist terminates the infinite series.

Table 6.1 shows calculated fundamental harmonic frequencies for a series of diatomics. Calculations were done at the SCF level with two different basis sets. We mentioned earlier the idea of a reference geometry. There is some disagreement amongst authors as to which reference geometry is appropriate for force constant calculation; the *experimental* geometry or that predicted at SCF level with the basis set in question.

Table 6.1 SCF calculations of ω_e (cm^{-1}) for some closed-shell first-row diatomics. In the TZVP case, force constants are given for the calculated and experimental geometries.

	STO/3G Calc. *R*	TZVP Calc. *R*	Expt. *R*	Expt.*
BeH^+	2508.97	2315.90	2298.32	1647.64
BeO	1939.18	1715.84	1557.49	1487.32
BH	2936.71	2507.01	2431.71	2366.
CH^+	3201.83	3081.68	3071.46	2739.54
HF	4473.91	4468.34	4172.93	4138.52
N_2	2666.92	2745.69	2515.88	2359.61

* Herzberg and Huber (1979)

Basis set dependence is not excessive for quadratic force constant calculations but an extended set is recommended. It is, however, apparent that one has to go beyond SCF theory in order to get quantitative agreement with experiment, despite the fact that such calculations are expensive. Correlation corrections are usually about 10% for single bonds and 20% for multiple bonds, and one can often allow for the effect of correlation by a suitable scaling.

6.2 POLYATOMIC MOLECULES

As stated in the previous section, diatomic molecules are very much a special case. The only diatomic method which can be used for a polyatomic molecule is the direct numerical point-by-point construction of the PE surface and subsequent numerical differentiations.

We saw in a previous chapter how molecular geometries can be predicted using gradient techniques. One calculates ∇U in order to determine a direction of search along the PE surface. Gradient techniques offer no advantage for a diatomic calculation, but for a polyatomic molecule the output contains $\partial U/\partial q_k$. So a general route to quadratic force constants is the direct *analytical* calculation of the gradient, followed by either an analytical or a *numerical* differentiation for $\partial^2 U/\partial q_i \partial q_j$. High numerical accuracy is attainable because the critical first differentiation is done analytically. Once again we should consider: (a) choice of basis set, (b) choice of reference geometry, and (c) representation of force constants in a suitable coordinate system. For points (a) and (b), the same comments apply as above. We recommend an extended basis set, e.g. STO/4-31G, and the *experimental* geometry if this is available.

Table 6.2 Basis set dependence of harmonic frequencies of water. All SCF calculations at the experimental geometry.

Basis		ω_e (cm^{-1})	
STO/3G	2045.6	4484.5	4785.8
STO/4-31G	1830.3	3883.6	3975.2
TZV	1799.2	3886.0	3973.1
TZVP	1786.8	3899.0	3982.6
Experiment	1648	3832	3943

Calculation of the PE gradient is best done in terms of nuclear Cartesian coordinates, and such force constants are perfectly acceptable for the calculation of vibrational frequencies. Chemists prefer to work with internal valence coordinates (bond stretching, bond angles, etc.) and it is usual to calculate U in Cartesian coordinates and then transform the results to internal valence coordinates. This is usually done automatically within the *ab initio* package.

Table 6.2 shows representative results at the SCF level for water. The agreement with experiment is impressive, but actually nowhere near the accuracy of spectroscopic measurements. Experience shows that the errors are largely systematic, and the conclusions of exhaustive studies are that: (a) diagonal stretching force constants are usually overestimated by 10–15%; (b) diagonal bending force constants are usually overestimated by 20% and for the off-diagonal constants the results are less systematic.

The most fruitful use of SCF force constants is in their combination with experimental data; the basic procedure is to scale the diagonal force constants and this leads to a better prediction of the off-diagonal terms which are themselves poorly determined experimentally.

CHAPTER 7

Electric Multipole Moments

So far we have been concerned with molecular energies and geometries. Molecular properties can be classified in various ways, and for our purposes we will divide them into properties which can be derived from the wavefunction of the electronic state under consideration, into properties that measure the response of a molecule to an applied electromagnetic field, and into spin properties.

Much of our knowledge of molecules is obtained by studying their response to electromagnetic radiation, and the very rapid recent growth in nonlinear spectroscopy and molecular electronics has focused attention on our ability to predict and rationalize the electric properties of molecules.

7.1 ELECTRIC DIPOLE MOMENTS

Suppose point charges $q_1, q_2, \ldots, q_n$ are located at positions $\mathbf{r}_1$, $\mathbf{r}_2$, $\ldots$, $\mathbf{r}_n$ respectively. The vector $\Sigma q_i \mathbf{r}_i$ is called the *electric dipole moment* and it is usually given the symbol $\mathbf{p}_e$:

$$\mathbf{p}_e = \Sigma q_i \mathbf{r}_i \tag{7.1}$$

The vector $\mathbf{p}_e$ has components $\Sigma q_i x_i$, $\Sigma q_i y_i$, $\Sigma q_i z_i$. The six quantities $q_i x_i^2$, $q_i y_i^2$, $q_i z_i^2, \ldots, q_i y_i z_i$ define the electric second moment $\mathbf{Q}$ which we can write as a 3×3 symmetric matrix

$$\mathbf{Q} = \begin{pmatrix} \Sigma q_i x_i^2 & \Sigma q_i x_i y_i & \Sigma q_i x_i z_i \\ \Sigma q_i x_i y_i & \Sigma q_i y_i^2 & \Sigma q_i y_i z_i \\ \Sigma q_i x_i z_i & \Sigma q_i y_i z_i & \Sigma q_i z_i^2 \end{pmatrix} \tag{7.2}$$

The collection of quantities $q_i x_i^3, \ldots, q_i x_i y_i z_i$ likewise defines the electric *third moments*, and so on. Many authors prefer to work with slightly different second, third, . . . moments, for reasons that will appear shortly. In particular, we are interested in the electric quadrupole moment $\boldsymbol{\Theta}$ defined by

$$\boldsymbol{\Theta} = \frac{1}{2}\begin{pmatrix} \Sigma q_i(3x_i^2 - r_i^2) & 3\Sigma q_i x_i y_i & 3\Sigma q_i x_i z_i \\ 3\Sigma q_i x_i y_i & \Sigma q_i(3y_i^2 - r_i^2) & 3\Sigma q_i y_i z_i \\ 3\Sigma q_i x_i z_i & 3\Sigma q_i y_i z_i & \Sigma q_i(3z_i^2 - r_i^2) \end{pmatrix} \tag{7.3}$$

If the charge distribution is spherically symmetrical then $\Sigma q_i x_i^2 = \Sigma q_i y_i^2 = \Sigma q_i z_i^2 = \frac{1}{3}\Sigma q_i r_i^2$ and so $\Theta_{xx} = \Theta_{yy} = \Theta_{zz} = 0$ and the electric quadrupole gives a measure of deviations from spherical symmetry. In any case.

$$(x_i^2 + y_i^2 + z_i^2) = r_i^2$$

so $\Theta_{xx} + \Theta_{yy} + \Theta_{zz} = 0$ always, and there are only five independent components of $\boldsymbol{\Theta}$.

Very similar considerations apply to the electric octupole, hexapole and decapole moments, although we do not often encounter any of these quantities in chemistry.

The quadrupole moment is an example of a *tensor* property. It always proves possible to find a set of coordinate axes which we label conventionally a, b, c such that

$$\boldsymbol{\Theta} = \begin{pmatrix} \Theta_{aa} & 0 & 0 \\ 0 & \Theta_{bb} & 0 \\ 0 & 0 & \Theta_{cc} \end{pmatrix} \tag{7.4}$$

and we refer to the values Θ_{aa}, Θ_{bb}, Θ_{cc} as the *principal values* of $\boldsymbol{\Theta}$ whilst the coordinate axes a, b, c are called the *principal axes*. If the charge distribution possesses axes of symmetry, these correspond to the principal axes.

7.2 THE MULTIPOLE EXPANSION

Why do we want to study electric moments? For a variety of reasons, most of which can be traced back to the multipole expansion of classical electromagnetism. The basic physical idea is as follows. Suppose we wish to calculate the electrostatic potential at point P in Figure 7.1. The electrostatic potential at $\mathbf{R}$ due to q_1 is

$$q_1/4\pi\varepsilon_0|(\mathbf{R} - \mathbf{r}_1)| \tag{7.5}$$

and so the total potential at $\mathbf{R}$ is

$$V(\mathbf{R}) = \sum \frac{q_i}{4\pi\varepsilon_0|\mathbf{R} - \mathbf{r}_i|} \tag{7.6}$$

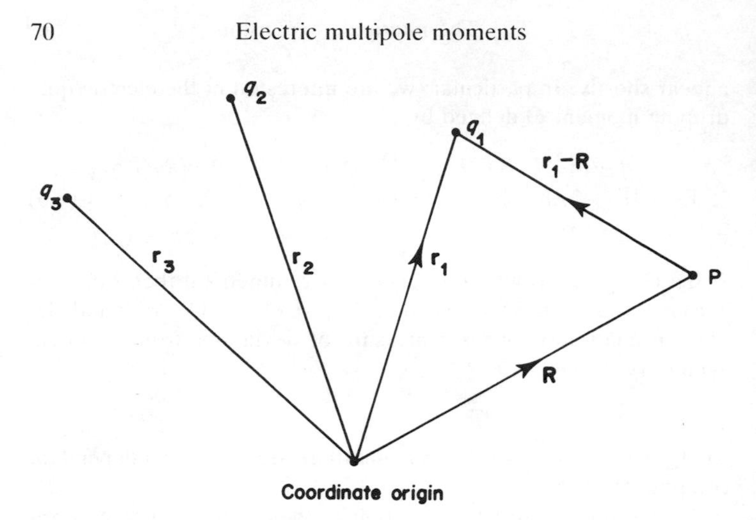

Figure 7.1 An array of point charges

If the point **R** is far from the array of charges, it may be more instructive to ask whether we can express the potential as some function of the general 'shape' of the charge distribution, and this is the notion of the multipole expansion. We expand the inverse distances about **R** as a Taylor series

$$\frac{1}{|\mathbf{R}-\mathbf{r}|}=\frac{1}{R}-\left(x\frac{\partial}{\partial X}(1/R)+y\frac{\partial}{\partial Y}(1/R)+z\frac{\partial}{\partial Z}(1/R)\right)$$
$$+\text{ higher terms}$$

and we recognize from an earlier chapter that the second term can be written as a gradient, thus

$$\frac{1}{|\mathbf{R}-\mathbf{r}|}=\frac{1}{R}-\mathbf{r}\cdot\boldsymbol{\nabla}(1/R)+\text{higher-order terms}$$

Obviously the higher-order terms will involve $x^2\partial^2/\partial X^2(1/R)$, etc. When we substitute this expression back into V we find

$$4\pi\varepsilon_0 V(\mathbf{R})=\Sigma q_i/R-(\Sigma q_i r_i)\cdot\nabla(1/R)+\ldots \tag{7.7}$$

i.e.

$$4\pi\varepsilon_0 V(\mathbf{R})=Q/R-\mathbf{p}_e\cdot\nabla(1/R)+\ldots \tag{7.8}$$

where Q is the total charge, $\mathbf{p}_e$ the electric dipole moment, etc. Each term on the right-hand side thus involves the product of a multipole moment and a term involving R^{-1}.

We will refer to the multipole expansion on several occasions in subsequent chapters. We simply state here for the record that the next extra term on the right-hand side of equation (7.8) is best written in terms of the electric field gradient tensor $\mathbf{E}'$

$$4\pi\varepsilon_0 V(\mathbf{R}) = Q/R - \mathbf{p}_e \cdot \nabla(1/R) + \Sigma\Sigma\, Q_{ij}E'_{ij} + \ldots \tag{7.9}$$

$$\mathbf{E}' = \begin{pmatrix} \dfrac{\partial E_X}{\partial X} & \dfrac{\partial E_X}{\partial Y} & \dfrac{\partial E_X}{\partial Z} \\ \dfrac{\partial E_Y}{\partial X} & \dfrac{\partial E_Y}{\partial Y} & \dfrac{\partial E_Y}{\partial Z} \\ \dfrac{\partial E_Z}{\partial X} & \dfrac{\partial E_Z}{\partial Y} & \dfrac{\partial E_Z}{\partial Z} \end{pmatrix} \tag{7.10}$$

and we will refer to this result later.

For a long time, the multipole expansion was widely regarded as the correct starting point for a study of intermolecular forces. It is interesting to note that all the classic texts on intermolecular forces begin with a discussion of electomagnetism and the multipole expansion. Historically, such studies were an indirect route to molecular quadrupole moments.

7.3 CHARGE DENSITIES

If instead of an array of point charges $q_1, \ldots, q_n$ we consider a continuous distribution of charge $\varrho(\mathbf{r})$, then the summations of equations (7.1), etc., have to be replaced by integrals:

$$\mathbf{p}_e = \int \varrho(\mathbf{r})\mathbf{r}\,d\tau \tag{7.11}$$

$$\Theta_{xx} = \tfrac{1}{2}\int (3x^2 - r^2)\varrho(\mathbf{r})\,d\tau \tag{7.12}$$

etc.

7.4 CALCULATIONS OF ELECTRIC DIPOLE MOMENTS

Electric dipole moments can be determined experimentally from the Stark effect, or from a study of dielectric polarization. The *sign* of the dipole moment of a linear molecule can be deduced from an isotope substitution study of the rotational molecular *magnetic* moment.

The electric dipole moment operator is

$$\hat{\mathbf{p}}_e = e\Sigma Z_\alpha \hat{\mathbf{R}}_\alpha - e\Sigma \hat{\mathbf{r}}_k \tag{7.13}$$

where the first sum runs over the nuclei and the second over the electrons. Z_α is the atomic number of nucleus α. Within the Born–

Oppenheimer approximation, we can therefore calculate the electric dipole moment

$$\mathbf{p}_e = e\Sigma Z_\alpha \mathbf{R}_\alpha - e\int \Sigma \hat{\mathbf{r}}_k |\Psi|^2 \mathrm{d}\tau \tag{7.14}$$

and the analogy between this equation and equations (7.1) and (7.11) should be apparent. $e|\Psi|^2$ is an *electron density*. Exactly the same comments apply to the calculation of higher moments

$$Q_{xx} = e\Sigma Z_\alpha X_\alpha^2 - e\int \Sigma \hat{\mathrm{x}}_k^2 |\Psi^2| \mathrm{d}\tau$$

etc. There will therefore be a nuclear and an electronic contribution to each property. The integrals are very simple. In each case, the expectation value is that of a sum of *one-electron operators* $\hat{\mathbf{r}}_k$, $\hat{\mathbf{x}}_k^2$, etc, and Brillouin's theorem encourages us to believe that the calculation of such properties for a closed-shell molecule should be reliable at SCF level. Brillouin's theorem states that single excited states constructed from Hartree–Fock orbitals do not interact with the electronic ground state:

$$\int \Psi_0^* \hat{H} \Psi_I^X \mathrm{d}\tau = 0$$

where Ψ_0 represents the SCF ground state wavefunction, and Ψ_I^X is a singly excited state wavefunction formed by promoting an electron from the Ith SCF orbital to the Xth virtual one.

Thus, if we attempted to use configuration interaction (CI) to improve the ground state wavefunction by mixing in singly, doubly, ...excited states,

$$\Psi = C_0\Psi_0 + \Sigma C_I^X \Psi_I^X + \Sigma\Sigma C_{IJ}^{XY} \Psi_{IJ}^{XY} + \ldots \tag{7.15}$$

we would calculate (e.g.) an electric dipole moment

$$\begin{aligned}\int \hat{\mathbf{p}}_e \Psi^2 \mathrm{d}\tau = {} & C_0^2 \int \hat{\mathbf{p}}_e \Psi_0^2 \mathrm{d}\tau + 2C_0 \Sigma C_I^X \int \hat{\mathbf{p}}_e \Psi_0 \Psi_I^X \mathrm{d}\tau \\ & + \Sigma\Sigma C_I^X C_J^Y \int \hat{\mathbf{p}}_e \Psi_I^X \Psi_J^Y \mathrm{d}\tau \\ & + \Sigma\Sigma C_{IJ}^{XY} C_{KL}^{UV} \int \hat{\mathbf{p}}_e \Psi_{IJ}^{XY} \Psi_{KL}^{UV} \mathrm{d}\tau + \ldots\end{aligned} \tag{7.16}$$

If we use perturbation theory to estimate the coefficients C_I^X, C_{IJ}^{XY}, ...then $C_I^X = 0$ according to Brillouin's theorem and the correction to the dipole moment is of the order of $(C_{IJ}^{XY})^2$ which explains the usual statement that SCF one-electron properties are correct to second order for closed-shell molecules.

As usual, we should enquire about: (a) the Born–Oppenheimer approximation, (b) basis set dependence, (c) correlation effects.

Breakdown of the Born–Oppenheimer approximation leads to a (tiny) nonzero electric dipole moment for CH_3D, but is otherwise thought to be totally unimportant to chemical accuracy.

Table 7.1 Effect of basis set on the electric dipole moment of pyridine.

Basis set	p_e (10^{-30} C m)	E/F_h
STO/3G	6.502	−243.628920
STO/4-31G	9.015	−246.318523
SV	9.464	−246.588409
TZV	9.080	−246.642356
TZVP	8.324	−246.738172
Experimental	7.31 ± 2%	

Basis set dependence is very important, and Table 7.1 shows the effect of basis set on the calculated electric dipole of pyridine. As a general rule, minimal basis sets underestimate electric dipole moments whilst extended basis sets overestimate them. Addition of polarization functions to extended basis sets usually *reduces* the calculated value.

Table 7.2 shows a representative sample of large basis set SCF calculations of electric dipole moment for molecules with closed shell electronic states. The agreement with experiment is in general very respectable.

SCF calculations on molecules with small electric dipole moments tend to be in poor absolute agreement with experiment. The oft-quoted example is CO, whose electric dipole moment has been deduced as 0.4076×10^{-30} C m. The polarity C^-O^+ was deduced from a molecular beam electric resonance experiment. Minimal basis sets give the correct polarity but extended basis sets give C^+O^-. The agreement with experiment is also usually poor for molecules like NH_3 which have a very low barrier to inversion (Table 7.3).

The error in SCF electric dipole moments is found quite uniformly to between 1 and 4×10^{-30} C m for closed-shell ground states of diatomic molecules and CI calculations have generally reduced the error by an order of magnitude. This apparent contradiction of the consequences of Brillouin's theorem has been thoroughly investigated in the special case of LiH and is easily explained: the bulk of the correction to the dipole is due to one single excitation from the highest occupied sigma (pi) orbital to the lowest unoccupied sigma (pi) orbital in molecules like LiH, HF, HCl and ClF.

It is also found that the SCF error in molecules with single bonds is only about one-third that of molecules with multiple σ or π bonds and this conclusion can be useful in predicting the expected accuracy of an SCF electric dipole calculation.

Table 7.2 SCF calculations of electric dipole moment for a selection of molecules. All calculations refer to extended, polarized basis sets.

	p_e (10^{-30} C m)			p_e (10^{-30} C m)	
Molecule	Calc.	Expt.*	Molecule	Calc.	Expt.
HF	−6.922	−6.093	HCl	−4.893	−3.60
HBr	−4.540	−2.70	HI	−3.459	−1.50
HCN	−11.052	−9.41	FCN	−7.540	−7.24
CLCN	−9.669	−9.41	BrCN	−10.675	−9.81
HCP	−1.340	−1.30	FCCH	3.918	2.4
ClCCH	2.104	1.5	CH_3F	−7.506	−6.23
CH_3Cl	−7.687	−5.94	CH_3Br	−7.488	−6.04
CH_3CN	−14.152	−13.05	SO_2F_2	−0.9402	−3.74
H_2O	−7.748	−6.17	H_2S	−4.857	−3.24
O	−8.127	−6.30	O	−2.899	−2.209
S	−8.671	−6.17	NH	−6.580	−6.14
H_2	1.637	1.40	S	−2.182	−1.328
=O	−12.154	−9.97	O	−7.767	−6.47
F	−6.941	−4.97	Cl	−6.977	−5.64
Br	−6.296	−5.67	N	−8.324	−7.31

Table 7.3 Calculated and experimental electric dipoles for CO, NH_3, PH_3 and AsH_3. The experimental value is shown in parenthesis.

Molecule	Basis set	p_e (10^{-30} C m)	
CO	STO/3G	0.5601	(0.4076)
	STO/6G	0.3406	
	STO/4-31G	−2.0104	
	TZVP	−1.0280	
NH_3	TZVP	6.496	(4.90)
PH_3	TZVP	3.291	(1.93)
AsH_3	TZVP	2.440	(0.67)

Table 7.4 Electric dipole moments for some open-shell diatomics.

		Dipole (10^{-30} C m)		
Molecule	State	SCF	CI	Expt.*
CH	$X\,^2\Pi$	13.3	12.1	12.4 ± 0.5
OH	$X\,^2\Pi$	15.1	13.8	14.1 ± 0.1
CN	$X\,^2\Sigma^+$	19.51	12.42	12.3 ± 0.7
CO^+	$X\,^3\Pi$	20.9	13.1	11.65
NO	$X\,^2\Pi$	2.20	−2.12	1.35
SO	$X\,^3\Sigma^-$	17.6	10.8	13.1 ± 0.2

* S. Green (1974).

Brillouin's theorem does not hold for open-shell electronic states, and the calculation of electric dipole moments is much less satisfactory in these cases. Some treatment of electron correlation is usually needed in order to obtain even modest agreement with experiment, and this is illustrated in Table 7.4.

As for closed-shell states, the size of the error is related to the availability of low-lying charge transfer excitations. If there are no low-lying charge transfer states (e.g. in the hydrides and highly ionic species), the SCF error is likely to be less than 3×10^{-30} C m. Otherwise the error may be very much larger. A small CI calculation containing only valence excitations reduces this error by an order of magnitude.

7.5 QUADRUPOLE MOMENTS

Electric quadrupole moments are not particularly easy to measure experimentally. Prior to 1970 the only direct routes to these quantities were from the Kerr and the Cotton–Mouton effects. They can now be obtained from molecular microwave Zeeman spectroscopy, to fair accuracy.

All the comments made about electric dipole moments also apply to the quadrupole moments. Table 7.5 shows the variation of quadrupole moment with basis set for pyridine. Apart from the minimal basis set calculation, all calculated values lie within the experimental error bars. Table 7.6 shows a selection of calculations for other molecules. The experimental quantities are not to be regarded as 'correct' any more than the *ab initio* values are 'correct'. Several of the experimental values refer to measurements corrected neither for the zero-point vibrations nor for the centrifugal effects.

Table 7.5 SCF calculations of the quadrupole moment of pyridine using five different basis sets.

	Θ (10^{-40} C m^2)		
Basis set	Θ_{aa}	Θ_{bb}	Θ_{cc}
STO/3G	−6.6219	20.0207	−13.3988
STO/4-31G	−19.4722	33.4409	−13.9687
SV	−21.3170	35.4157	−14.0987
TZV	−21.3170	34.6943	−13.5189
TZVP	−23.1643	33.9562	−10.792
Experimental	−21 ± 5	33 ± 4	−12 ± 3

Table 7.6 Representative electric quadrupole moment calculations.

	Θ_{zz} (10^{-40} C m^2)	
Molecule	Calc.	Expt.
KF	−25.8261	−31.3 ± 2.3
CO	−7.6500	−6.7 ± 3.3
CO^+	−19.1654	−14. ± 1.
HCN	7.9282	10. ± 2.
FCCH	15.0354	13.2 ± 0.5
PH_3	−9.402	−7.0 ± 3.3
CH_3F	−1.133	−4.7 ± 3.7
CH_3CN	−8.606	−6.0 ± 4.0
 S	−1.7684	−1.7 ± 2.3
	+6.9765	4.0 ± 2.7
	−5.2081	−2.3 ± 2.3
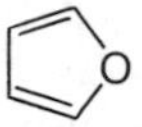 O	0.063	0.7 ± 1.3
	22.717	19.7 ± 1.0
	−22.780	−20.4 ± 1.3
 S	4.550	5.7 ± 5.3
	23.195	22.0 ± 5.0
	−27.746	−27.7 ± 7.3
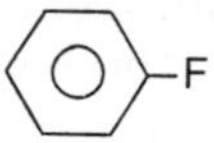 F	−8.216	−6.4 ± 2.7
	29.550	17.0 ± 3.3
	−21.306	−10.6 ± 3.3

Table 7.7 Higher electric moments of CH_4. *xyz* component of the octupole moment and *xxxx* component of the hexadecapole.

Comment	*xyz* (ea_0^3)	*xxxx* (ea_0^4)
'Experiment'	3.23 ± 0.60	−12.1 ± 1.5
Extended spd	2.9271	−7.8390
Extended spd +extra diffuse spd	2.5363	−7.8209
Very large basis set	2.48	−7.90

A general conclusion is that *ab initio* SCF calculations offer a reliable route to the electric quadrupole tensor.

7.6 HIGHER ELECTRIC MOMENTS

Electric moments beyond the second are rarely encountered in chemistry. They do appear in the field of collision-induced spectroscopy, and Table 7.7 shows a comparison with experiment. There is no particular reason to assume that the experimental values are any more reliable than the SCF ones.

CHAPTER 8

Electric Field Gradients

We have discussed some of the consequences of electron spin in earlier chapters. You should recall that electrons possess a non-zero internal angular momentum which we call 'spin', and we write this angular momentum vector **S**. Electron spin obeys the usual quantum-mechanical rules of angular momentum, and because electrons are charged particles they are also magnetic dipoles. It is possible to measure simultaneously the size of the vector **S** and of any one (but *only* one) of its components which we label conventionally the z component. The size of **S** is $\sqrt{S(S+1)}\,\hbar$ and of the z component $m_s\hbar$, where $S = 1/2$ and $m_s = \pm S$.

Many *nuclei* have a corresponding internal angular momentum which we refer to as *nuclear spin* and we use the symbol **I** to represent the vector. The spin quantum number I is characteristic of the given nucleus and can have different values for different isotopic species. Many nuclei with $I \geqslant 1$ also possess a quadrupole moment which is generally written eQ and defined without the factor 1/2 as in Chapter 7. Thus for example,

$$Q_{zz} = \int \varrho_N (3z^2 - r^2)\mathrm{d}\tau \tag{8.1}$$

where ϱ_N is the *nuclear* charge distribution and the integral is over the nuclear coordinates. Nuclear quadrupole moments have to be determined experimentally.

The charged particles in the nucleus can be thought of as rotating very rapidly about the z-axis of the nuclear spin vector, and if an average is made over a time long enough for the nuclear particles to rotate but so short that the electrons have not appreciably changed position, the nuclear charge distribution may be considered cylindrical. Using a principal axis system which coincides with the nuclear spin, all non-diagonal components of Q are zero and $Q_{xx} = Q_{yy} = -\frac{1}{2}Q_{zz}$. The entire quadrupole tensor is expressed in terms of 'the' value $Q = Q_{zz}$.

In a molecule, a given nucleus will generally experience an electric field gradient and simple arguments show that only p and d atomic orbitals can contribute to field gradients. An electric quadrupole $\mathbf{\Theta}$ has energy of interaction

$$W = \tfrac{1}{2}\Sigma\Sigma\Theta_{ij}E'_{ij} \tag{8.2}$$

with a field gradient $\mathbf{E}'$. Remembering the e and the $\frac{1}{2}$, the corresponding expression for a nuclear quadrupole is

$$W = -\tfrac{1}{6}e\Sigma\Sigma Q_{ij}q_{ij} \tag{8.3}$$

where we have used the conventional symbol $\mathbf{q}$ for the electric field gradient at the nucleus. In principal axes, the interaction is characterized by Q_{aa} and two of q_{aa}, q_{bb}, q_{cc}. The *largest* of q_{aa}, q_{bb}, q_{cc} is given the symbol q, and we refer to eqQ/h as the *quadrupole coupling constant.* The quantity η = (difference of smaller q's/largest of the q's) is called the asymmetry parameter.

8.1 EXPERIMENTAL DETERMINATION

Quadrupole coupling constants for molecules are usually determined from the hyperfine structure of pure rotational spectra or from electric beam and magnetic beam resonance spectroscopies. Nuclear magnetic resonance, electron spin resonance, and Mossbauer spectroscopies are also routes to the property.

A large amount of experimental data exists for ^{14}N and halogen substituted molecules. There is less available data for deuterium because the nuclear quadrupole is very small.

8.2 CALCULATIONS OF eQq/h

All we have to do is to evaluate the electric field gradient at each nucleus. Like electric dipoles and quadrupoles, this is a straightforward calculation. The electric field at point P of Figure 7.1 is given by

$$\mathbf{E} = \frac{1}{4\pi\varepsilon_0}\sum_i \frac{q_i}{|\mathbf{R} - \mathbf{r}_i|^3}(\mathbf{R} - \mathbf{r}_i) \tag{8.4}$$

For a molecular calculation we need to generalize this to a contribution from the nuclei which we treat as point charges, and from the electron density $\varrho(\mathbf{r})$,

$$\mathbf{E} = \frac{e}{4\pi\varepsilon_0}\left\{\sum_\alpha \frac{Z_\alpha}{|\mathbf{R} - \mathbf{R}_\alpha|^3} - \int \frac{|\Psi|^2(\mathbf{R} - \mathbf{r})}{|\mathbf{R} - \mathbf{r}|^3}\,\mathrm{d}\tau\right\} \tag{8.5}$$

Table 8.1 SCF calculation of eQq/h for N_2.

Basis set	TZVP
R_e	109.4 pm
E/E_h	−108.978 047
Contributions to field gradient q_{zz} at nucleus	
Nuclear	1.5844
Electron	−2.8886
Total	−1.3043
Q	0.016×10^{-28} m^2
eQq/h	−4.903 MHz

Table 8.2 Calculated ^{14}N quadrupole coupling constants at SCF level for pyridine.

Basis set	eQq/h (MHz)	Asymmetry η
Experimental	−4.88 ± 0.04	0.405 ± 0.005
STO/3G	−4.4367	0.3677
STO/4-31G	−4.7553	0.2879
SV	−5.0416	0.2134
TZV	−5.0573	0.2730
TZVP	−4.7738	0.3297

We calculate the field gradient simply as $E'_{xx} = \partial E_x/\partial x$, etc., and this is usually done internally within the *ab initio* package. Performing the calculation for N_2 at R = 109.4 pm gives the data in Table 8.1. The experimental value is 4.648 MHz. The sign is a more elusive quantity to evaluate experimentally.

Table 8.2 shows the variation with basis set at SCF level for a representative larger molecule, pyridine. As a general rule, minimal basis sets give a rather poor representation of q and it is wise to stick to extended sets. Asymmetry factors are very elusive animals.

The same comments obviously apply for nuclei other than ^{14}N. There is, for example, some data on deuterium and a large amount on halogen quadrupole coupling in the literature. Deuterium data is hard to obtain experimentally because the the nuclear quadrupole is small, and because hydrogen atoms in molecules are almost spherical.

CHAPTER 9

The Electrostatic Potential

9.1 INTRODUCTION

One of the fundamental objectives of chemistry is the prediction and rationalization of molecular reactivity. In principle, this involves the calculation of a potential energy surface in the first instance. We noted in an ealier chapter the labour and expense involved in constructing *ab initio* potential energy surfaces to chemical accuracy for even simple reactions, and it is probably fair to say in summary that such predictions have so far failed to materialize, for everyday reactions.

Most of the traditional theories of chemical reactivity have been aimed at organic molecules, and these theories usually attempt to extract from the electronic properties of an isolated molecule, some useful information as to how the molecule will interact with other molecules.

We distinguish between *static* theories, which in essence give a description of the electronic wavefunction, and *dynamic* theories, which aim to predict, for example, the response of a molecule to an approaching point charge. The 'classical' approaches are concerned with calculations of bond order, free valency, autopolarizability, etc.

In recent years, the electrostatic potential has been used to give a rapid static representation of the outstanding features of molecular reactivity, so it is appropriate to revise our ideas about potentials at this point.

The electrostatic *field* $\mathbf{E}(\mathbf{R})$ at position $\mathbf{R}$ due to the point charge q_1 at position $\mathbf{r}_1$ is

$$\mathbf{E}(\mathbf{R}) = \frac{q_1}{4\pi\varepsilon_0}\frac{(\mathbf{R} - \mathbf{r}_1)}{|\mathbf{R} - \mathbf{r}_1|^3} \tag{9.1}$$

If we move a 'test charge' Q_0 from position $\mathbf{R}_A$ to position $\mathbf{R}_B$ in this static field, it is easy to demonstrate that the work done:

(i) is independent of path;
(ii) can be written $Q_0(V_B - V_A)$;

where the electrostatic potential V_B at point $\mathbf{R}_B$ is given by

$$V_B = \frac{q_1}{4\pi\varepsilon_0} \cdot \frac{1}{|\mathbf{R}_B - \mathbf{r}_1|} \tag{9.2}$$

$Q_0 V_B$ represents the work done in bringing Q_0 from infinity to $\mathbf{R}_B$. The quantity $Q_0 q_1/4\pi\varepsilon_0|\mathbf{R}_B - \mathbf{r}_1|$ is called the *mutual potential energy* of Q_0 and q_1, and the definitions above can be generalized as follows. For an array of point charges $q_1, q_2, \ldots, q_n$ at positions $\mathbf{r}_1, \mathbf{r}_2, \ldots, \mathbf{r}_n$, the mutual potential energy U is

$$U = \frac{1}{4\pi\varepsilon_0} \sum\sum_{i<j} \frac{q_i q_j}{|\mathbf{r}_i - \mathbf{r}_j|} \tag{9.3}$$

and this represents the work done in building up the charge distribution. The potential energy at position $\mathbf{R}$ is

$$V(\mathbf{R}) = \frac{1}{4\pi\varepsilon_0} \sum \frac{q_i}{|\mathbf{R} - \mathbf{r}_i|} \tag{9.4}$$

whilst for a continuous charge distribution $\varrho(\mathbf{r})$ we write

$$V(\mathbf{R}) = \frac{1}{4\pi\varepsilon_0} \int \frac{\varrho(\mathbf{r})}{|\mathbf{R} - \mathbf{r}|} \mathrm{d}\tau \tag{9.5}$$

In the special case of a molecule, which we regard as a set of point charges (the nuclei) and a continous electron density, we write

$$4\pi\varepsilon_0 V(\mathbf{R}) = \sum \frac{Z_\alpha e}{|\mathbf{R} - \mathbf{R}_\alpha|} - e \int \frac{|\Psi|^2 \mathrm{d}\tau}{|\mathbf{R} - \mathbf{r}|} \tag{9.6}$$

Once we have used quantum mechanics to calculate $\varrho(\mathbf{r})$, the expression for $V(\mathbf{R})$ is exactly what we would get from classical electromagnetism. The *electronic* contribution is a sum of one-electron operators, and all the considerations of previous chapters apply.

One generally presents calculations of the electrostatic potential as contour diagrams, and to make our discussion concrete we give several simple examples. It is usual to present contour diagrams of $+1\mathrm{C}\ V(\mathbf{R})$, i.e. the work done in bringing a unit positive charge to position $\mathbf{R}$. We will, however, refer to such maps as potential energy maps.

9.2 WATER MOLECULE

The spirit of this kind of calculations is to give a rough description of molecular reactivity, and so we do not intend to pay any attention to basis set dependence or correlation effects. All the figures presented are STO/3G calculations.

Figures 9.1 and 9.2 show such contour maps for H_2O in either symmetry plane. The maps give work done moving a *proton* from infinity to **R**. Such maps therefore give directly the interaction energy of the molecule with an electrophile. Potential values are obviously high in the region of the nuclei, and it is usual to ignore

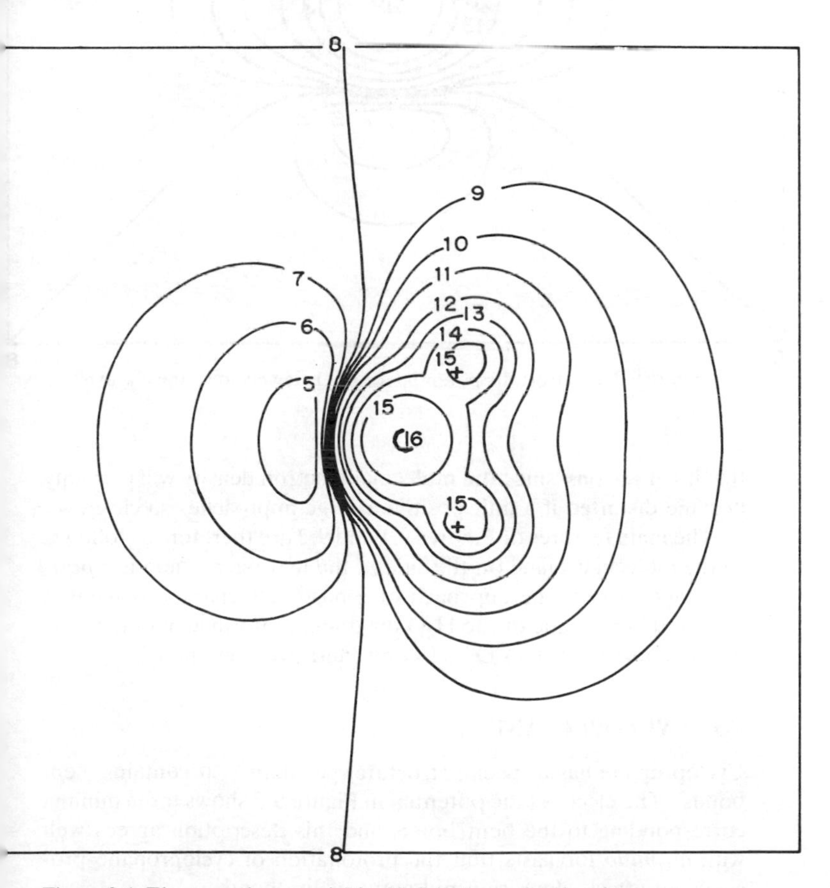

Figure 9.1 Electrostatic potential for H_2O in the molecular plane

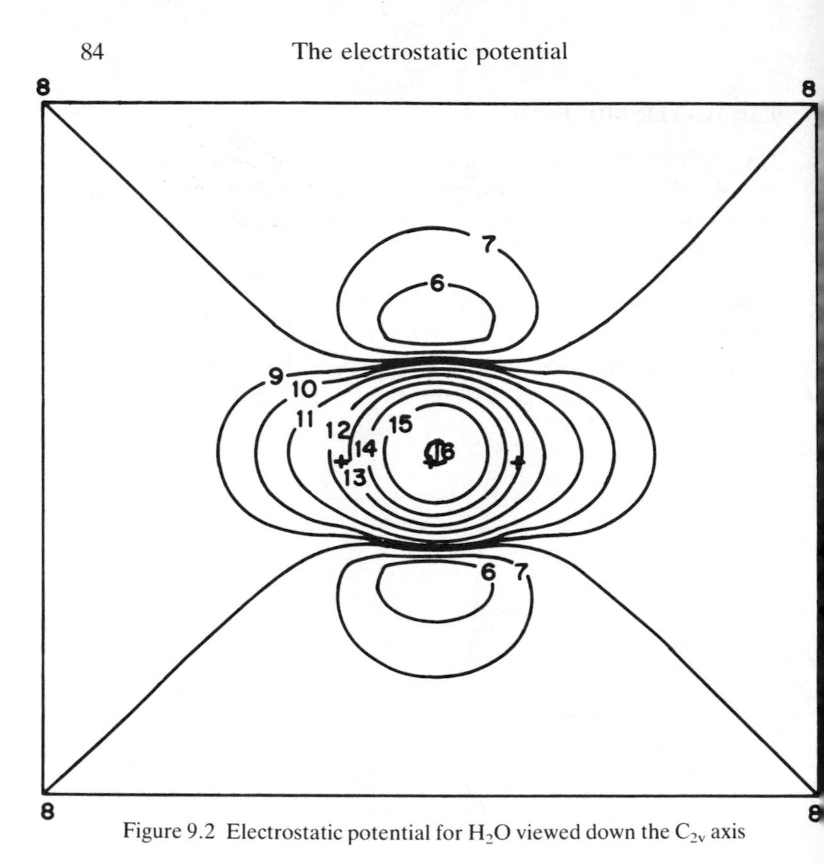

Figure 9.2 Electrostatic potential for H_2O viewed down the C_{2v} axis

the 'inner regions' since the molecular electron density will certainly become distorted if a unit positive charge approaches so close.

The main features of Figures 9.1 and 9.2 are therefore as follows. In the molecular plane on the side of the hydrogen, there is a *positive* region where the approach of a positively charged reagent is favoured. The shape of the H_2O diagrams conform generally to the intuitive picture of two O—H bonds and two 'lone pairs'.

9.3 CYCLOPROPANE

Cyclopropane has a special structural peculiarity; it contains 'bent bonds'. The electrostatic potential in Figure 9.3 shows three minima corresponding to the bent bonds and this description agrees well with *ab initio* forecasts that the protonation of cyclopropane proceeds by attack along the mid-point of the bond.

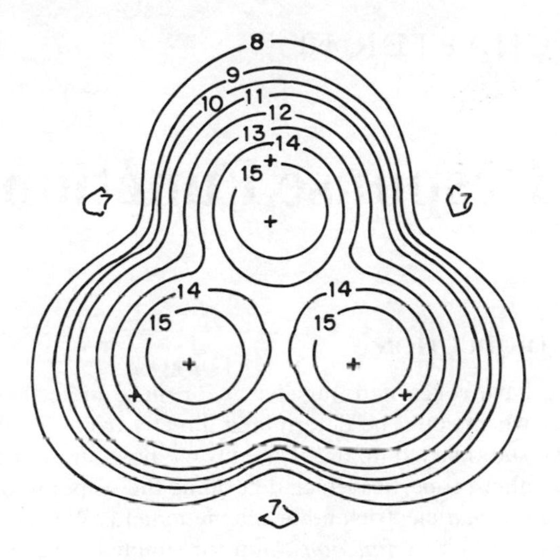

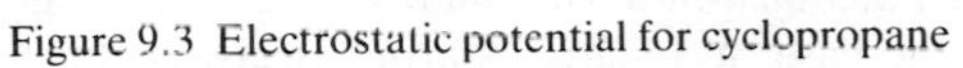
Figure 9.3 Electrostatic potential for cyclopropane

CHAPTER 10

Response Functions

10.1 INTRODUCTION

So far we have concerned ourselves with those molecular electric properties which could be determined directly from the wavefunction of the stationary state under study. We now turn our attention to certain other properties which determine the response of a molecule to an *external* electric and/or magnetic field. We refer to such properties as *response functions*, and for simplicity we restrict the discussion initially to the case of a static external electric field.

10.2 A STATIC EXTERNAL ELECTRIC FIELD

In Chapter 1 we saw how the multipole expansion permitted the electrostatic potential V due to an arbitrary charge distribution to be written in terms of the electric moments and a characteristic distance R from the charge distribution. Thus,

$$V(\mathbf{R}) = \{Q/R - \mathbf{p}_e \cdot \nabla(1/R) + \ldots\} \tag{10.1}$$

where Q is the total charge, $\mathbf{p}_e$ the electric dipole moment, etc.

If we now switch on an external electrostatic field $\mathbf{E}$, this field may well induce a dipole moment in addition to any permanent dipole moment and there is no particular reason to assume that the direction of the induced dipole will necessarily be in the direction of the applied field. Thus for example, we write for the x component of $\mathbf{p}_e$,

$$p_{e,x}(\mathbf{E}) = p_{e,x}(\mathbf{0}) + \sum_{i=x,y,z} \alpha_{xi} E_i + 1/2! \sum_{i,j=x,y,z}\sum \beta_{xij} E_i E_j + \ldots \tag{10.2}$$

where the sums run over the x, y and z components of the electric dipole vector (etc.). There is obviously an analogous expression for the y and z components, and we can write the equation much more succinctly in tensor notation as

$$\mathbf{p}_e(\mathbf{E}) = \mathbf{p}_e(0) + \boldsymbol{\alpha}\cdot\mathbf{E} + \tfrac{1}{2}\mathbf{E}\cdot\boldsymbol{\beta}\cdot\mathbf{E} + \ldots \quad (10.3)$$

The quantities α_{ij} define the electric dipole polarizability tensor, and the tensor $\boldsymbol{\beta}$ is called the first dipole hyperpolarizability.

The dipole polarizability tensor is symmetric, $\alpha_{ij} = \alpha_{ji}$, and so there are no more than six independent components. As with all tensor properties, it is usual to seek a set of molecular axes such that the tensor is diagonal:

$$\boldsymbol{\alpha} = \begin{pmatrix} \alpha_{aa} & 0 & 0 \\ 0 & \alpha_{bb} & 0 \\ 0 & 0 & \alpha_{cc} \end{pmatrix} \quad (10.4)$$

and for a molecule with symmetry, these principal axes correspond to the molecular axes. We are very often interested in the mean polarizability $\bar{\alpha} = \frac{1}{3}(\alpha_{aa} + \alpha_{bb} + \alpha_{cc})$. For a linear molecule there are only two independent components which we write $\alpha_{\parallel}$, $\alpha_{\perp}$ and we define the anisotropy $\beta = \alpha_{\parallel} - \alpha_{\perp}$. β is usually a positive quantity.

Molecules generally respond to a *dynamic* electric field in a quite different way than to a static field, and reference is usually made to static and dynamic polarizabilities. Polarizability is said to be a frequency-dependent phenomenon, and some care has to be exercised when comparing theoretical and experimental values, because different values will result when measuring with different frequencies.

10.3 A STATIC EXTERNAL MAGNETIC INDUCTION

We have had very little to say so far about magnetic properties. For the most part, electric properties of molecules are easy to understand because the particles responsible for electric effects are electrons and protons. Once a given electron distribution has been calculated quantum mechanically, the usual equations of classical electomagnetism apply.

Obviously, Maxwell's equations apply at the molecular level, but magnetic phenomena are somewhat different in that there appear (probably) to be no fundamental magnetic 'monopoles', the analogues of elementary charge particles. At least, if such particles do exist they have never been detected so far!

We remark here that the atomic phenomenon responsible for magnetism is angular momentum. Any charged particle with non-zero angular momentum $\mathbf{M}$ is a magnetic dipole, and we write the magnetic dipole moment

$$\mathbf{p}_{\mathrm{M}} = g\frac{Q}{2m}\mathbf{M} \tag{10.5}$$

where Q is the charge, m the mass and g an experimental factor called the Landé g-factor. We often write this as $\mathbf{p}_{\mathrm{M}} = \gamma\mathbf{M}$ where γ is the gyromagnetic ratio.

Magnetic poles beyond the dipoles are rarely encountered in chemistry, and the effect of an external magnetic induction $\mathbf{B}$ on a magnetic dipole turns out to be adequately represented by

$$\mathbf{p}_{\mathrm{M}}(\mathbf{B}) = \mathbf{p}_{\mathrm{M}}(\mathbf{B}{=}0) + \boldsymbol{\varkappa}\cdot\mathbf{B} \tag{10.6}$$

$\boldsymbol{\varkappa}$ is called the *magnetizability*, and it is the magnetic analogue of the polarizability.

In order to make progress with the quantum-mechanical treatment, we need to enquire how the Hamiltonian $\hat{H}_0$ for a molecule is affected by external electric and magnetic fields, and we now attend to this.

10.4 THE HAMILTONIAN

We introduced earlier the concept of the electrostatic potential, $V(\mathbf{R})$, which is related to the electric field $\mathbf{E}(\mathbf{R})$ by $\mathbf{E} = -\boldsymbol{\nabla}V$. When discussing magnetic phenomena, it turns out that it is not in general useful to introduce such a simple potential, rather one has to work with a more complicated construct called the vector potential $\mathbf{A}$, defined by $\mathbf{B} = \boldsymbol{\nabla} \times \mathbf{A}$ where $\mathbf{B}$ is the magnetic induction.

Consider a particle of mass m and charge q moving in the external static fields $\mathbf{E}$ and $\mathbf{B}$, where $\mathbf{E} = -\boldsymbol{\nabla}V$ and $\mathbf{B} = \boldsymbol{\nabla} \times \mathbf{A}$. The field-free Hamiltonian is

$$\hat{H}_0 = \frac{p^2}{2m} \tag{10.7}$$

and it turns out that the correct Hamiltonian in the fields is given by

$$\hat{H} = \frac{(\mathbf{p} - q\mathbf{A})^2}{2m} + qV \tag{10.8}$$

the operator $\mathbf{p} - q\mathbf{A}$ is called a generalized momentum. In the Schrödinger picture, $\mathbf{p} \rightarrow -i\hbar\boldsymbol{\nabla}$ and we find eventually

$$\hat{H} = \hat{H}_0 + \frac{i\hbar q}{m}\mathbf{A}\cdot\boldsymbol{\nabla} + \frac{q^2A^2}{2m} + qV \tag{10.9}$$

In the case of a uniform electrostatic field along the z axis, $V = -Ez$ where the zero of potential is taken as the coordinate origin, giving

$$\hat{H} = \hat{H}_0 - qEz$$

or more generally

$$\hat{H} = \hat{H}_0 - \mathbf{p}_e \cdot \mathbf{E} \tag{10.10}$$

Conversely, in the special case of a uniform magnetic induction with no electric field, where an induction B along the z axis is

$$\mathbf{A} = -\tfrac{1}{2}\mathrm{B}(\mathbf{i}y - \mathbf{j}x)$$

we find

$$H = H_0 + \frac{i\hbar qB}{2\mathrm{m}}\left(-y\frac{\partial}{\partial x} + x\frac{\partial}{\partial y}\right) + \frac{q^2B^2}{8m}(x^2 + y^2) \tag{10.11}$$

provided **A** satisfies $\nabla \cdot \mathbf{A} = 0$.
From the definition of angular momentum, we can rewrite this as

$$\hat{H} = \hat{H}_0 - \frac{qB}{2m}\hat{l}_z + \frac{q^2B^2}{8m}(x^2 + y^2) \tag{10.12}$$

We will refer to each expression in the text.

10.5 CALCULATION OF DIPOLE POLARIZABILITY

We saw earlier that

$$\mathbf{p}_e(\mathbf{E}) = \mathbf{p}_e(\mathbf{E}{=}0) + \boldsymbol{\alpha} \cdot \mathbf{E} + \ldots$$

and that the energy W of a charge distribution in the presence of an external electrostatic field could be written

$$W = QV - \mathbf{p}_e \cdot \mathbf{E} - \tfrac{1}{3}\Sigma\,\Sigma\,\Theta_{ij}E'_{ij} \tag{10.13}$$

If the charge distribution is mobile, it will redistribute itself until W is minimized and the electric dipole moment will therefore change according to

$$W = W_0 = \mathbf{p}_e(\mathbf{0}) \cdot - \tfrac{1}{2}\mathbf{E} \cdot \boldsymbol{\alpha} \cdot \mathbf{E} \tag{10.14}$$

This gives us two routes to polarizability: we can calculate the induced dipole or the second-order energy. Unlike electric dipoles and the other properties we have studied so far, the dipole polarizability measures the *response* of a molecule to an external electric field and it cannot be written as the expectation value of a sum of one-electron operators. There are essentially two options open for polarizability calculations, and we discuss each in turn.

10.5.1 Perturbation theory

Suppose that $\hat{H}_0$ is the Hamiltonian for the molecule in the absence of an applied field. $\hat{H}_0$ could for example be a Hartree–Fock Hamiltonian which we had already determined from an SCF calculation.

In the spirit of perturbation theory, we could treat $-\mathbf{p}_e \cdot \mathbf{E}$ as a perturbation on a solved problem, and we could calculate $\boldsymbol{\alpha}$ according to equation (10.14) as a second-order contribution to the energy. For the sake of argument, assume that the state of interest is the ground state, and so

$$W^{(2)} = -\sum_{m>0} \frac{|\int \Psi_m^* (-\mathbf{p}_e \cdot \mathbf{E}) \Psi_0 \, d\tau|^2}{E_m - E_0} \tag{10.15}$$

giving for example

$$\alpha_{xx} = 2 \sum_{m>0} \frac{|\int \Psi_m^* \hat{p}_{e,x} \Psi_0 \, d\tau|^2}{E_m - E_0} \tag{10.16}$$

where $p_{e,x}$ is the x component of the dipole moment operator. This equation is rather unhelpful for practical calculation, because it requires a knowledge of all the excited states Ψ_m. Such knowledge is hard to come by!

Let us return to SCF calculations. Suppose we have calculated an LCAO SCF wavefunction for the molecule of interest. This means that we have determined a set of column vectors $\mathbf{C}_A, \mathbf{C}_B, \ldots, \mathbf{C}_M$ (assuming a closed-shell molecule with M electron pairs) which satisfy the matrix eigenvalue problem

$$\mathbf{h}^F \mathbf{C}_K = \mathbf{E}_K \mathbf{S} \mathbf{C}_K \tag{10.17}$$

and are mutually orthogonal in the sense that

$$\mathbf{C}_K \mathbf{S} \mathbf{C}_L = \delta_{KL} \tag{10.18}$$

The matrix R_0 defined by

$$\mathbf{R}_0 = \Sigma \, \mathbf{C}_K \mathbf{C}_K^+ \tag{10.19}$$

where $\mathbf{C}_k^+$ is the Hermitian transpose of $\mathbf{C}^k$, $\mathbf{C}_k$ is often referred to as the *electron density matrix* and it is easily shown that

$$\mathbf{R}_0 \mathbf{h}^F = \mathbf{h}^F \mathbf{R}_0$$

and

$$\mathbf{R}_0 \mathbf{S} \mathbf{R}_0 = \mathbf{R}_0 \tag{10.20}$$

These two equations are really restatements of the fact that the $\mathbf{C}_K$'s satisfy the eigenvalue problem above, and are orthogonal.

If we now add a perturbation $\boldsymbol{\Delta}$ to $\mathbf{h}^{\mathrm{F}}$, the electron density (and $\mathbf{h}^{\mathrm{F}}$, which is defined in terms of $\mathbf{R}$) will now change and we can develop a perturbation expansion

$$\mathbf{R}_0 \rightarrow \mathbf{R}_0 + \mathbf{R}^{(1)} + \ldots$$

such that equation (10.20) is still satisfied. A self-consistent solution is necessary, and the technique is referred to as self-consistent perturbation theory (SCPT). The advantage of this approach is that the orders of perturbation theory are rigorously separated; the polarizability is calculated directly from the second-order energy expression.

10.5.2 Finite fields

By far the more usual approach is to add the perturbation directly to the Hamiltonian, and redo the variational calculation. In the case of an SCF calculation we simply add $\boldsymbol{\Delta}$ to the Hartree–Fock Hamiltonian matrix and solve the new SCF equation

$$(\mathbf{h}^{\mathrm{F}} + \boldsymbol{\Delta})\mathbf{C}_K = e_K \mathbf{S}\mathbf{C}_K$$

This involves calculating a matrix of dipole integrals over atomic orbitals and choosing a finite value for the applied field E. The size of E must be sufficiently large as to give acceptable accuracy in the polarizability, which is usually calculated from the induced dipole, and sufficiently small to ensure that second-order effects in $\mathbf{p}_e$ are negligible. Typically one uses E = 0.001 a.u.

This approach is called the finite field (FF) method, and it was first proposed by Cohen and Roothann (1965).

10.5.3 An example

An example should make everything clear. Table 10.1 shows a SCPT calculation for HF using a triple-zeta basis set augmented with polarization functions. In the SCPT case we calculate directly the second-order energy $W^{(2)}$ and obtain α from $= -W^{(2)}/E^2$. In this simple case we know the direction of the principal axes of α; in the case of a molecule with lower symmetry it would be necessary to calculate *all* elements of α by switching on the field in three separate and mutually perpendicular directions.

Before getting round to comparisons with experiment, we must remark that the calculation of polarizability is extremely basis-set-

Table 10.1 Self-consistent perturbation calculation of the dipole polarizability of HF.

	Geometry	$R = 1.6978\ a_0$	
	Basis set	Dunning TZVP	
	SCF energy	$-100.060\,696\ E_h$	
	Permanent dipole	$-0.81716\ ea_0$	
	External field	0.001 a.u.	
	parallel	perpendicular	
Second-order energy	−0.76735	−0.20999	$\times 10^{-6}\ E_h$
α	1.535	4.200	$e^2a_0{}^2E_h^{-1}$

dependent, as is evident from Table 10.2. A calculation of the polarizability of hydrogen using simply a 1*s* orbital would give zero; at least *p* orbitals would need to be added in order to give the required response to the field.

For much the same reason, the STO/3G basis set calculation of HF (Table 10.2) like all minimal basis set calculations is essentially worthless for calculation of $\alpha_\perp$. This is because the basis set is not sufficiently flexible to respond to the applied field.

$\alpha_\parallel$ is moderately well represented at the STO/3G level. Moving to STO/4-31G gives a slight improvement all round, and is starting to look more healthy. The TZVP-1 calculation includes polarization functions on both H and F, with standard exponents, chosen from energy optimization studies on small molecules. The exponents were $\alpha_p(\mathrm{H}) = 1.0$, $\alpha_d(\mathrm{F}) = 1.62$. Energy optimization studies tend to neglect the outer regions of the molecules which also happen to be the most easily polarized. In the calculation labelled TZVP-2,

Table 10.2 Calculated dipole polarizability for HF using a variety of basis sets.

		$\alpha\ (e^2a_0{}^2E_h{}^{-1})$	
Basis set	E/E_h	$\alpha_\perp$	$\alpha_\parallel$
STO/3G	−98.568 128	0.002	2.988
STO/4-31G	−99.886 632	0.640	3.618
TZVP-1	−100.060 696	1.535	4.200
TZVP-2	−100.043 365	3.945	5.311
TZVP-3	−100.055 658	4.154	5.458

Table 10.3 Calculated and experimental polarizabilities for some simple molecules. SCF calculations using TZVP-4 basis sets.

Molecule			$\bar{\alpha}^0$		β^0 (10^{-40} C^2 m^2 J^{-1})	
	$\alpha_{\parallel}^0$	$\alpha_{\perp}^0$	Calc.	Exp.	Calc.	Exp.
N_2	2.4574	1.5884	1.8781	1.935	0.8690	0.77
CO	2.3885	1.8463	2.0270	2.20	0.5423	0.59
NNO	4.9050	2.0605	3.0087	3.37	2.8445	3.28
CO	3.9240	1.9719	2.6225	3.241	1.9523	2.34
OCS	8.0854	4.0973	5.4267	5.80	3.9881	4.58
CH_4	2.6414	2.6414	2.6414	2.885	0	0
HCCH	5.1477	3.1518	3.8171	4.3	1.9959	2.07

we have used *polarizability-optimized* polarization functions. The exponents used were $\alpha_p(\mathrm{H}) = 0.2$ and $\alpha_d(\mathrm{F}) = 0.209$. There is an immediate improvement in $\alpha_\perp$ and a small change in $\alpha_\parallel$.

If we wish to make further progress, we have to increase the size of the basis set, and in particular it is necessary to include more polarization functions. In calculation TZVP-3, we have used a double set of polarization functions with exponents $\alpha_p(\mathrm{H}) = 0.2$, 0.5 and $\alpha_d(\mathrm{F}) = 0.139$, 0.418. As a general rule, the 'best' single d exponent for a first-row atom is given by $\eta_0 = 0.0034\,(Z - 1.16)^2$ where Z is the atomic number. The best double set of d orbital exponents is then $2/3\eta_0$, $2\eta_0$ and finally if one wishes to extend the set further it is usual to add a d orbital set with exponent $8\eta_0$, which is usually close to the energy-optimized value (for fluorine, $8\eta_0 = 1.672$).

Table 10.3 shows the level of accuracy to be expected, even when using very large basis sets. The experimental quantities are $\bar{\alpha}^0$ and β^0.

10.6 CALCULATIONS OF MAGNETIZABILITY

We saw in an earlier section that in the case of an external magnetostatic induction,

$$\hat{H} = \hat{H}_0 - \frac{qB}{2m}\hat{L}_z + \frac{q^2B^2}{8m}(x^2 + y^2) \tag{10.21}$$

provided that the vector potential $\mathbf{A}$ satisfies $\nabla \cdot \mathbf{A} = 0$. If we use perturbation theory to calculate the first- and second-order contributions to energy we find

$$\Delta W^{(1)} = \int \Psi_0^* \left(-\frac{qB}{2m} \hat{L}_z \right) \Psi_0 \, d\tau \tag{10.22}$$

$$\Delta W^{(2)} = \frac{q^2 b^2}{8m} \int \Psi_0^* (\hat{x}^2 + \hat{y}^2) \Psi_0 \, d\tau$$

$$- B^2 \sum_{m>0} \frac{\left\{ \int \Psi_m^* \left(-\frac{qL_z}{2m} \right) \Psi_0 \, d\tau \right\}^2}{E_m - E_0} \tag{10.23}$$

The magnetizability is given by the sum of a *diamagnetic* term

$$\varkappa^{d}_{zz} = - \frac{q^2}{4m} \int \Psi_0^* (x^2 + y^2) \Psi_0 \, d\tau \tag{10.24}$$

and a *paramagnetic* term

$$\varkappa^{p}_{zz} = 2 \sum_{m>0} \frac{\left\{ \int \Psi_m \left(-\frac{qL_z}{2m} \right) \Psi_0 \, d\tau \right\}^2}{E_m - E_0} \tag{10.25}$$

$\varkappa^d$ and $\varkappa^p$ are both tensor properties, and the remaining elements can be found by cyclic permutation of x, y and z. In the case of a molecule, the Hamiltonian contains extra terms to take account of shielding (a given nucleus will not see *exactly* an applied field because of the presence of nearby magnetic nuclei) and we return to this effect in a later section.

In the magnetic case, there is a very simple relationship between the magnetizability $\varkappa$ (a molecular property) and the susceptibility χ (a bulk property). The molar susceptibility $\chi_M = \mu_0 N\varkappa/V$ where N is the number density and V the volume of the sample. The terms magnetizability and susceptibility tend to be used interchangeably and we will follow this usual practice.

In the *electric* case, the relationship between polarizability (molecular phenomen) and polarization (bulk phenomenon) is by no means straightforward.

10.6.1 The diamagnetic contribution

The diamagnetic contribution is very easy to calculate, and comes out as the expectation value of the second-moment operators discussed in an earlier chapter. All the remarks made in that chapter hold, and the agreement with experiment is usually excellent, even with very simple basis sets (e.g. STO/3G). Table 10.4 shows a representative sample for four cyclic compounds.

Table 10.4 Representative SCF calculations of χ^d using a TZVP basis set.

		χ^d (10^{-5} J T^{-2} mol^{-1})	
		Calc.	Expt.
Furan	xx	−317.2	−313.9
	yy	−188.3	−189.5
	zz	−186.8	−182.5
Benzene		−291.2	−286
		−291.2	−286
		−509.6	−508
Pyridine (N)		−733.3	−732.4
		−515.4	−509.7
		−296.6	−293.3
Cyclopentadiene (H_2)		−358.1	−356.9
		−214.6	−213.0
		−216.6	−214.8

10.6.2 The paramagnetic contribution

Although calculation of the paramagnetic contribution is formally the same as the calculation of electric dipole polarizability, there are two major differences as far as the practical calculation is concerned. If we rewrite the Hamiltonian in terms of the vector potential **A**,

$$\hat{H} = \hat{H}_0 + \frac{i\hbar q}{m}\mathbf{A}\cdot\boldsymbol{\nabla} + \frac{q^2A^2}{2m}$$

we see that the perturbation is imaginary and it is necessary to work with complex wavefunctions.

Secondly we note that the definition

$$\mathbf{B} = \boldsymbol{\nabla} \times \mathbf{A}$$

leaves the vector potential undetermined, because we can add $\boldsymbol{\nabla}\phi$ to **A** where ϕ is any differentiable scalar field, and still find

$$\mathbf{B} = \boldsymbol{\nabla} \times (\mathbf{A} + \boldsymbol{\nabla}\phi) = \boldsymbol{\nabla} \times \mathbf{A}$$

Obviously, any calculated property should be independent of this *choice of gauge*, but unfortunately this is not the case in finite basis set SCF calculations. One way to solve the problem is to work with so-called gauge-invariant atomic orbitals, which are atomic orbitals of the form $\phi_k \exp(\gamma \mathbf{A}_k \cdot \mathbf{r})$ where ϕ_k is a (usual) atomic orbital centred on nucleus k and $\mathbf{A}_k$ the vector potential at that nucleus.

10.7 MAGNETIC SHIELDING

Many fundamental particles have an internal angular momentum called 'spin'. In particular, protons have a spin for which the spin quantum number is 1/2. In the presence of an external magnetic induction $\mathbf{B}$, a proton with magnetic moment $\mathbf{p}_{\mathrm{M}} = \alpha\mathbf{I}$ has an energy of interaction $-\mathbf{p}_{\mathrm{M}} \cdot \mathbf{B} = -\alpha\mathbf{I} \cdot \mathbf{B}$, and this forms the basis for the field of nuclear magnetic resonance (nmr) spectroscopy.

If the proton is embedded in a molecule, the magnetic induction the proton sees is not exactly the applied field, because surrounding magnetic dipoles also generate their own fields, and we write

$$\mathbf{B}_{\mathrm{eff}} = \mathbf{B}_{\mathrm{ext}}(1 - \sigma)$$

where σ is the magnetic shielding tensor. In high-resolution nmr experiments the components of σ are molecule fixed, and one really measures the mean value $\bar{\sigma} = \frac{1}{3}(\sigma_{aa} + \sigma_{bb} + \sigma_{cc})$ where a, b and c refer to the principal axes. To calculate $\bar{\sigma}$ we have to consider the effect on the molecule of the external field *and* a magnetic dipole, and this can be done by using the vector potential

$$\mathbf{A} = \frac{1}{2}(\mathbf{B} \times \mathbf{r}) + \frac{\mu_0}{4\pi} \frac{\mathbf{p}_{\mathrm{M}} \times \mathbf{r}}{r^3}$$

The chemical shift is then calculated from the energy.

Just as in the theory of diagmagnetic susceptibility, the chemical shift has a diamagnetic part σ^{d} and a paramagnetic part σ^{p}. Exactly the same methods as used for χ are used to evaluate σ^{d} and σ^{p}.

CHAPTER 11

Recovering Chemical Concepts

11.1 INTRODUCTION

The era of 'modern' quantum chemistry began in 1926 with the work of Schrödinger, of Heisenberg and of Dirac. Dirac's famous statement (the subject of many final-year BSc Chemical Physics examination questions!) summarized the state of the art for 1929.

> The underlying physical laws necessary for the mathematical theory of a large part of physics and the whole of chemistry are thus completely known and the difficulty is only that the exact application of these laws leads to equations much too complicated to be soluble.

What Dirac meant was that the technology was not at that time available for detailed solutions of the basic equations (at least, that is what I think he meant!)

It was not until the 1950s that things began to change. Consider these two statements:

> The hazy paths of electrons whirling round a nucleus and the structure and reactions of complex molecules are practical computing assignments for automatic calculating machines.
>
> *Time Magazine*, 23 January 1950

and again:

> It has thus been established that the only difficulty which exists in the evaluation of the energy and the wavefunction of any molecule...is the amount of computing necessary.
>
> S. F. Boys, 1950

Even after reading this book you will probably take exception to the following remark:

> We can calculate everything.
>
> E. Clementi and C. A. Coulson, 1973

but we have obviously come a long way since 1926.

So far we have concerned ourselves with calculations of physical observables: energies, electric dipole moments and the like. These quantities are of great interest to certain types of experimentalist, but at the end of the day we have to ask how the study of such quantities helps our understanding of chemical concepts (and more importantly, how a knowledge of chemical concepts helps us to understand the values of certain physical quantities). A study of the index of Pauling's classic book *The Nature of the Chemical Bond* (1939) reveals a list of concepts such as Bent single bonds, Bond energy, Covalent radii, Double bond character, Electronegativity, ..., Unshared electron pairs.

Chemistry is essentially the science of small changes among similar atoms and molecules, and we really need to be able to identify from the mathematics concepts such as the ones above. It is fair comment to say that this study has only been recently resumed. In this final chapter we will therefore consider the following goal:

> ...to give insights, not numbers.
>
> E. Clementi and C. A. Coulson, 1973

11.2 ORBITALS AND CHARGE DENSITY

The result of an SCF LCAO calculation is a set of molecular orbitals, and an obvious question is 'What do they look like?'. Most users of *ab initio* packages will have access to graphics packages such as GINO and GHOST, and there are two normal ways of

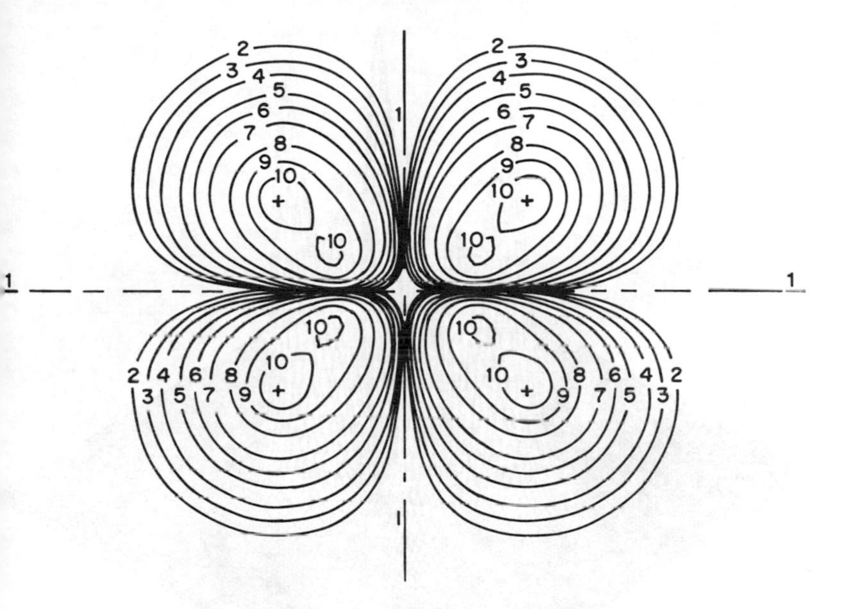

Figure 11.1 Contour diagram of a σ-orbital of ethene. The quantity plotted is the square of the wavefunction

presenting orbitals graphically. Figure 11.1 shows a *contour diagram* for the highest-energy occupied σ-orbital of ethene. An alternative approach is to give a three-dimensional representation, and this can be done in several ways. Figure 11.2 is a particular representation of the same orbital as a three-dimensional view called an *isometric projection* (the orbital is taken to be at an infinite distance from the viewpoint, and the three coordinate axes make equal angles with the direction of view. In an isometric projection, distances marked along the axes appear at equal intervals. In a *perspective* projection, the viewpoint is taken to be at a finite distance from the orbital, and so 'parallel' lines get closer together as the distance from the viewpoint increases.)

An obvious problem in plotting individual orbitals is that there are a lot of them. Many theories of organic reactivity involve a detailed consideration of the shape and nodal properties of the highest occupied and lowest unoccupied orbitals (the HOMOs and the LUMOs), but we should note that molecular orbitals determined by the SCF procedure are not unique. You probably remember from the theory of determinants that if

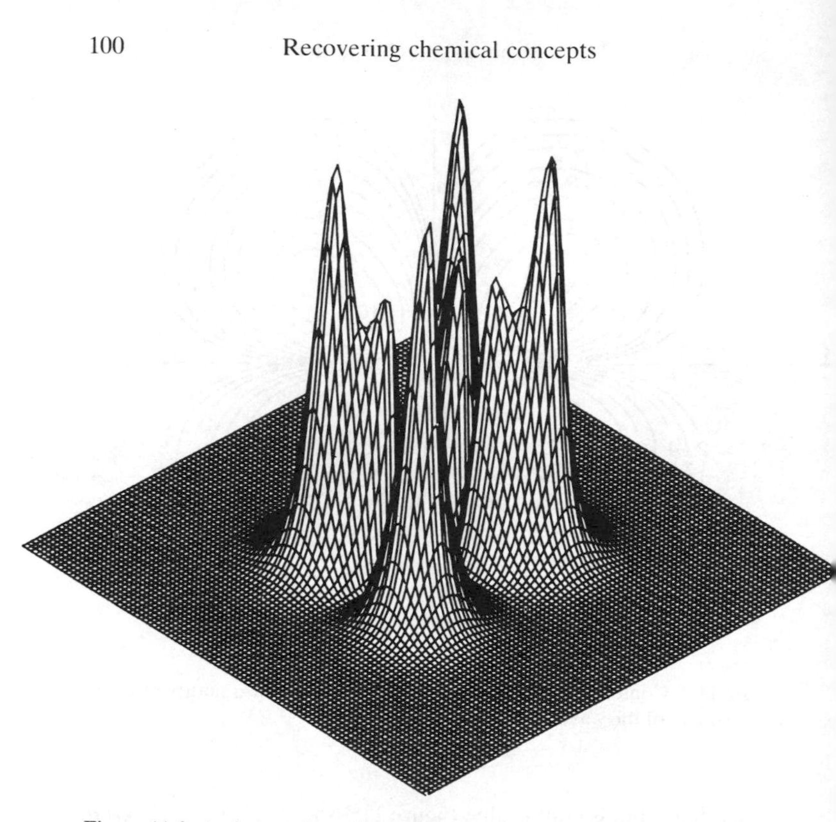

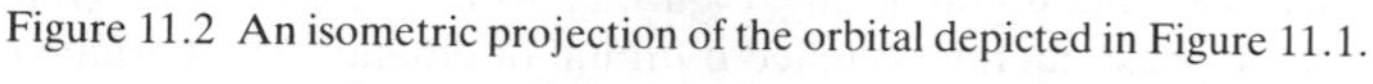

Figure 11.2 An isometric projection of the orbital depicted in Figure 11.1.

Again, the quantity plotted is the square of the wavefunction

$$D_1 = \begin{vmatrix} 1 & 4 & 7 \\ 2 & 5 & 8 \\ 3 & 6 & 9 \end{vmatrix} \qquad D_2 = \begin{vmatrix} 1 & 5 & 7 \\ 2 & 7 & 8 \\ 3 & 9 & 9 \end{vmatrix}$$

then $D_1 = D_2$ because the second column of D_2 is equal to the first column of D_1 plus the second column of D_1. Thus, going back to Chapter 3, we noted that the generalized Pauli principle forced us to use Slater determinants as the building blocks of electronic wavefunctions. A helium $1s^1 2s^1$ orbital configuration would be represented as

$$\Psi = N_1 \begin{vmatrix} 1s(\mathbf{x}_1) & 1s(\mathbf{x}_2) \\ 2s(\mathbf{x}_1) & 2s(\mathbf{x}_2) \end{vmatrix} \tag{11.1}$$

which we could also write in terms of the linear combinations $\chi_+ = 1s + 2s$; $\chi_- = 1s - 2s$ as

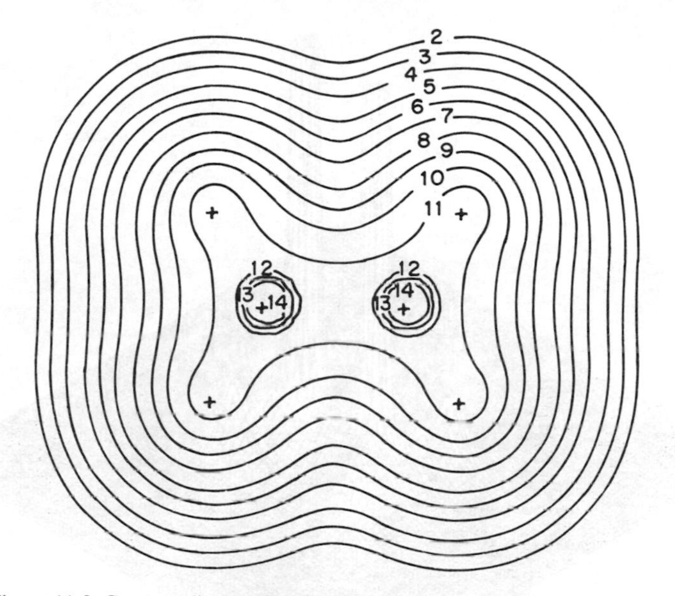

Figure 11.3 Contour diagram for the total electron density in ethene

$$\Psi = N_2 \begin{vmatrix} \chi_+(\mathbf{x}_1) & \chi_+(\mathbf{x}_2) \\ \chi_-(\mathbf{x}_1) & \chi_-(\mathbf{x}_2) \end{vmatrix} \tag{11.2}$$

where we have assigned electrons 1 and 2 to the new orbitals ($1s$ + $2s$) and ($1s - 2s$).

In any case, wavefunctions by themselves have no direct physical interpretation. According to the Born interpretation, we interpret $|\Psi|^2 \mathrm{d}V$ as a probability. For a many-electron wavefunction, $|\Psi|^2 \mathrm{d}V$ becomes $|\Psi|^2(\mathbf{x}_1, \mathbf{x}_2, \ldots, \mathbf{x}_n)\mathrm{d}\mathbf{x}_1\mathrm{d}\mathbf{x}_2 \ldots \mathrm{d}\mathbf{x}_n$, which gives us a probability for the instantaneous configuration of all the electrons. Many common physical properties depend only on the probability per unit volume of finding a *single* electron (no matter which) at a given point $\mathbf{r}$ in space, and this is given by

$$P(\mathbf{r}) = n \int |\Psi|^2 \mathrm{d}s_1 \mathrm{d}\mathbf{r}_2 \mathrm{d}s_2 \ \ldots \ \mathrm{d}\mathbf{r}_n \mathrm{d}s_n \tag{11.3}$$

The factor n arises because the n electrons are indistinguishable, and after integration we replace the variable $\mathbf{r}_1$ by $\mathbf{r}$. The function $P(\mathbf{r})$ is often referred to as the 'electron density', since for many purposes the electron distribution may be treated as a smeared-out charge of density P (electrons per unit volume). Figures 11.3 and 11.4 show a contour plot and an isometric projection for the total

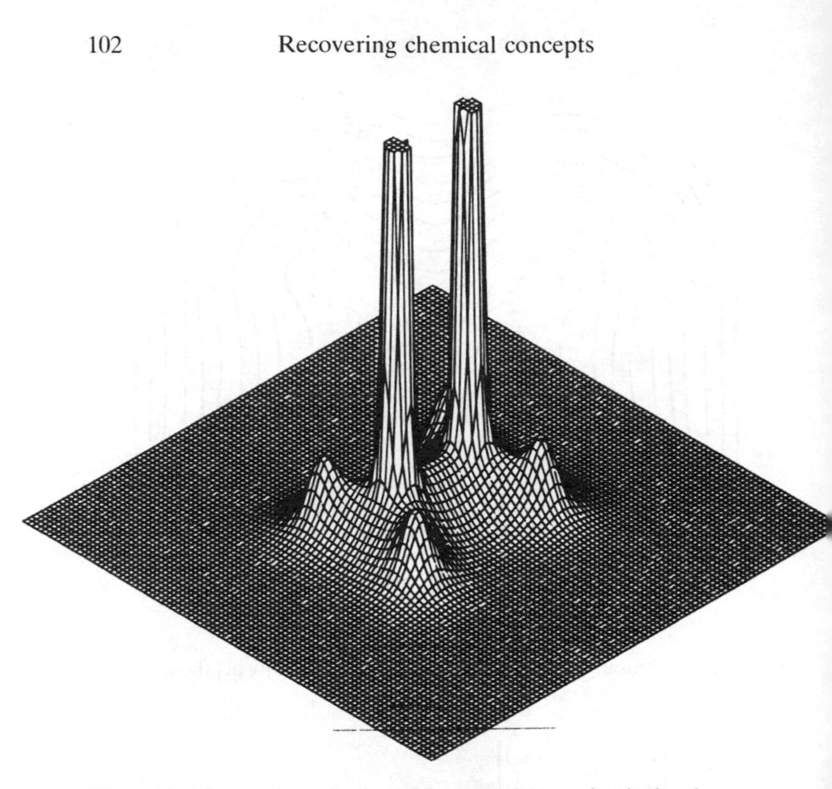

Figure 11.4 Isometric projection of the total electron density in ethene

electron comparable to the electron density maps deduced from X-ray crystallography, and the 1970s saw a deal of activity in this field.

11.3 LOCALIZED ORBITALS

By their very nature, LCAO MOs are delocalized over large regions of a molecule (as in Figure 11.1). When G. N. Lewis first proposed his classical theory of chemical bonding, the theory was dominated by concepts such as inner shells, lone pairs and directed bonds, and chemists traditionally think along Lewis's lines.

How do we recover this traditional picture from an *ab initio* calculation based on delocalized orbitals? We have already answered the question in Section 11.2; we make use of the arbitrariness of a determinantal wavefunction and impose an additional constraint in order to recover a localized picture.

Several constraints are possible. Most chemists have been educated on a diet of VSEPR (valence shell electron pair repulsion) theory, and an obvious constraint is to use these ideas in order to require that each electron pair is as far away in space from every other. Electric dipole moments give a simple measure of the distribution of a charge density, so the strategy of our localization method (actually due to S. F. Boys, and so usually referred to as *Boys Localization*), is:

(i) define a dipole for each electron pair;
(ii) take linear combinations of occupied orbitals in such a way that the repulsion between all electron pairs is minimized (i.e. all electron pairs are as far away from each other as possible.)

The calculation is an iterative one; we seek a set of LCAO MOs $\omega_1, \omega_2, \ldots, \omega_M$ which are doubly occupied and are given as linear combinations of the SCF LCAO MOs, such that

$$D = \sum_{i=1}^{M} \{\int \omega_i^*(\mathbf{r})\mathbf{r}\omega_i(\mathbf{r})\mathrm{d}\tau\}^2 \qquad (11.4)$$

is a maximum. Each integral in the sum is readily identifiable as a *bond dipole*, so the Boys technique seeks to maximize the sums of squares of the distances of orbital centroids from a molecular coordinate origin. This is equivalent to maximizing the sums of squares of the distances between orbital centroids, and to minimizing the repulsion between electron pairs. The method is particularly easy to program, and inexpensive in computer time.

Figures 11.5–7 show sample contour plots for the localized or-

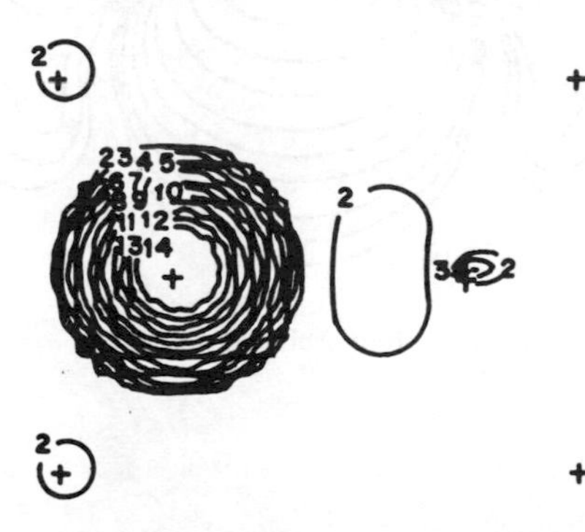

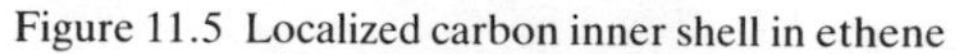
Figure 11.5 Localized carbon inner shell in ethene

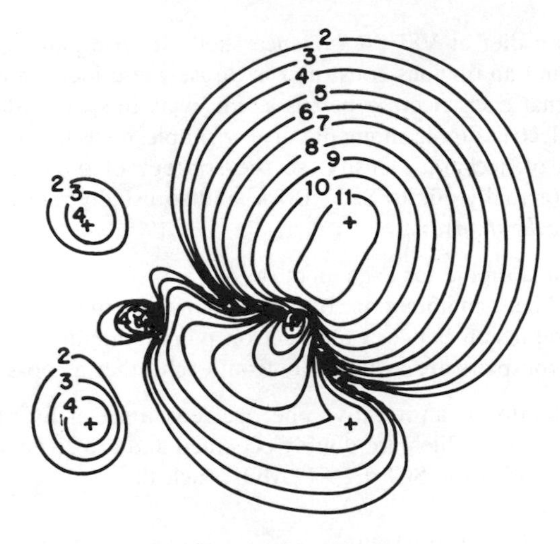

Figure 11.6 Localized C—H bond orbital in ethene

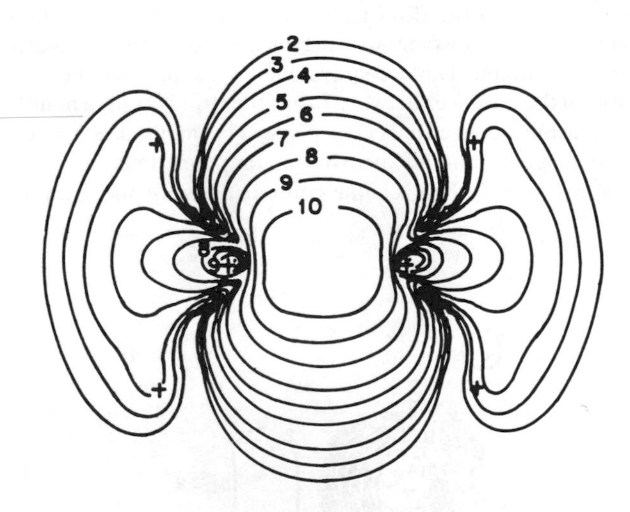

Figure 11.7 Localized C—C bond orbital in ethene

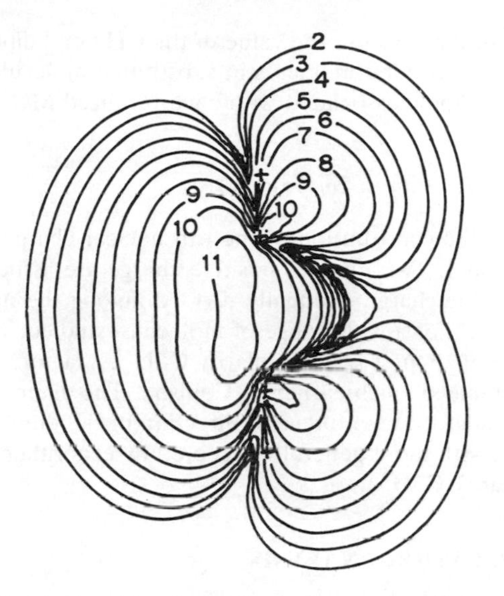

Figure 11.8 Banana bond in ethene. The hydrogen atoms lie in a plane perpendicular to the plane of the paper

bitals in ethene. Instead of the delocalized SCF MOs we recover the traditional chemical pictures of bonds, inner shells, etc. An interesting feature of such calculations for unsaturated compounds is that we often obtain 'banana bonds' rather than the usual picture of σ/π separation (see Figure 11.8).

11.4 BOND PROPERTIES

Chemistry is dominated by the bond concept, and a large part of predictive physical chemistry uses the idea that a given molecular property such as an enthalpy of formation or an electric dipole moment can be calculated as a sum of contributions from bonds (and lone pairs and inner shells). A number of investigations have been made at the *ab initio* level into bond additivity and transferability. Almost all such investigations start from localized orbitals, and for illustration we tell now the (unfinished) story of the CH bond dipole.

There has been a certain controversy about the value and the polarity of the CH bond dipole, arguments having been put forward for both C^+H^- and C^-H^+ polarities with an absolute value of between 0.4 D (1 D = 3.34×10^{-30} C m) and 1.0 D. Such knowledge is highly desirable because the value of the CH bond dipole is used to set the scale of bond moments in substituted molecules.

The bond dipole associated with each localized MO ω_i is given by

$$\int \mathbf{r}\{\Sigma Z_{i\alpha}\delta(\mathbf{R}_\alpha - \mathbf{r}) - 2\omega_i^*(\mathbf{r})\omega_i(\mathbf{r})\}\mathrm{d}\tau$$

where $Z_{i\alpha}$ is the contribution of the αth nuclear charge to the ith localized orbital. Two units of positive charge are associated with each ω_i and the charge is equally divided among the nuclei associated with ω_i. For a wide range of molecules studied, we found a bond dipole of about 2 D with polarity C^+H^-, a factor of 2 greater than that obtained much earlier by Coulson. The dilemma remains that the calculated magnitude and direction of the dipole are quite inconsistent with those generally accepted by experimentalists (0.4 D with polarity C^-H^+).

11.5 POPULATION ANALYSIS

The aim of population analysis is to divide up molecules into 'atoms' and 'overlap regions', and to characterize these regions numerically by the amount of charge they contain. If we are dealing with an SCF calculation where the LCAO orbitals $\psi_A, \psi_B, \ldots, \psi_M$ are occupied by $\nu_A, \nu_B, \ldots, \nu_M$ electrons (where the occupation number ν is 1 or 2), then the charge density is simply the sum of the individual orbital charge densities:

$$P(\mathbf{r}) = \nu_A\psi_A^*(\mathbf{r})\psi_A(\mathbf{r}) + \ldots + \nu_M\psi_M^*(\mathbf{r})\psi_M(\mathbf{r}) \qquad (11.5)$$

i.e. $P(\mathbf{r}) = \Sigma \nu_R\psi_R^*(\mathbf{r})\psi_R(\mathbf{r})$

Each LCAO MO will have been expressed in terms of LCAO coefficients and an atomic orbital basis set $\phi_1, \phi_2, \ldots, \phi_n$, so we can write

$$P(\mathbf{r}) = \Sigma\Sigma\,\phi_i^*(\mathbf{r})\mathbf{P}_{ij}\phi_j(\mathbf{r}) \qquad (11.6)$$

where $\mathbf{P}$ is called the 'charges and bond orders' matrix beloved by exponents of empirical π-electron theories.

In population analysis we divide up the electron density $P(\mathbf{r})$ into 'atomic' and 'overlap' populations

$$P(\mathbf{r}) = \Sigma\, P^A(\mathbf{r}) + \tfrac{1}{2}\Sigma\Sigma\, P^{AB}(\mathbf{r})$$

where the first sum runs over atoms A and the second over pairs of atoms A, B. $P^A(\mathbf{r})$ is called the *net density* of atom A, and $P^{AB}(\mathbf{r})$ is the *overlap density* associated with atoms A and B. The density $P^{A}(\mathbf{r})$ is calculated from an expression like (11.6) above, but the sum over atomic orbitals is taken only over orbitals centred on atom A. The overlap density P^{AB} involves orbitals centred on atom A and orbitals centred on atom B. To obtain a numerical measure of the electron density in each region we integrate to get the corresponding *populations*:

$$p^{A} = \int P^{A}(\mathbf{r})d\tau$$
$$p^{AB} = \int P^{AB}(\mathbf{r})d\tau$$

These quantities can be easily calculated from the LCAO MO coefficients and the atomic orbital overlap integrals.

There are three basic assumptions of population analysis:

(i) the partitioning described above is chemically useful;
(ii) useful information is retained when the densities are integrated to give populations;
(iii) it is valid to divide the overlap populations into parts and assign the parts to the two contributing atoms

$$p_{AB} = \nu_{A}{}^{AB} + \nu_{B}{}^{AB} \qquad (11.7)$$

For atom A, the sum of all the fractions $\nu_{A}{}^{AB}$, $\nu_{A}{}^{AC}$, ..., $\nu_{A}{}^{AZ}$ is called the valence population ν_{A}. The sum of ν_{A} and the atom population p_{A} is called the *gross population* q_{A}.

The Mulliken scheme is the most widely used, and the fractions in equation (11.7) are assumed equal. This is obviously unfair for a molecule such as LiF, but the general belief in population analysis is that a comparison of indices over a related series of molecules will lead to 'chemically reasonable' answers.

Population analysis indices are particularly easy to calculate, but they are markedly basis set dependent. Table 11.1 shows Mulliken indices for ethene calculated using: (i) a minimal STO/3G basis set,

Table 11.1 Mulliken population analysis indices for ethene calculated using a minimal basis set (STO/3G) and an extended basis set.

	STO/3G	Extended
p^{C}	6.1291	6.3091
p^{H}	0.9354	0.8454
p^{CH}	0.7908	0.8456
p^{CC}	1.1850	1.2277

and (ii) an extended, polarized set. The moral of this calculation is clear; it is only meaningful to compare population analysis indices for calculations with very similar basis sets. Unfortunately this was not realized by the early workers!

Population analysis tends to be used in two main ways, and the relative importance of assumptions (i), (ii) and (iii) above is different in each case. Often we take the 'atomic charges' $Z^{A} - q^{A}$, where Z^{A} is the nuclear charge, as a succinct description of the charge distribution and polarity. When population analysis is used in this way, approximation (i) is less important than (ii). The second main use of population analysis indices is in a discussion of the nature and strength of chemical bonding, with p^{AB} being taken as a measure of the strength of a bond.

References

R. F. W. Bader and R. A. Gangi, in *Specialist Periodical Reports: Theoretical Chemistry vol 2*, R. N. Dixon and C. Thomson (eds), Royal Society of Chemistry, London, 1976

W. S. Benedict and E. K. Plyler, *Can. J. Phys.*, **35** (1957), 1235

M. P. Bogaard and B. J. Orr, in International Review of Science, *Physical Chemistry Series 2 vol 2*, A. D. Buckingham (ed), London, Butterworths, 1975

M. Born and R. Oppenheimer, *Ann. Phys.*, **84** (1927), 457

S. F. Boys and F. Bernardi, *Molec. Phys.*, **19** (1970), 553

S. F. Boys, *Rev. Mod. Phys.*, **32** (1960), 306

S. F. Boys, *Proc. Roy. Soc. London.*, **A200** (1950), 542

M. L. Brillouin, *Actual. Sci. Ind.*, **71** (1933)

E. Clementi and C. A. Coulson, *Proceedings of the Robert A. Welch Foundation Conferences on Chemical Research XVI*, Robert A. Welch Foundation, Houston, 1973, pp. 61 and 117

E. Clementi and D. L. Raimondi, *J. Chem. Phys.*, **38** (1963), 2686

H. D. Cohen and C. C. J. Roothaan, *J. Chem. Phys.*, **43** (1965), S34

J. B. Collins, P. van R. Schleyer, J. S. Binkley and J. A. Pople, *J. Chem. Phys.*, **64** (1976), 5142

D. J. DeFrees, B. A. Levi, S. K. Pollack, W. J. Hehre, J. S. Binkley and J. A. Pople, *J. Am. Chem. Soc.*, **101** (1979), 4085

B. Dickinson, *J. Chem. Phys.*, **1** (1933), 317

P. A. M. Dirac, *Proc. Roy. Soc. London.*, **A123** (1929), 714

J. L. Dunham, *Phys. Rev.*, **41** (1932), 721

T. H. Dunning and P. J. Hay, in *Modern Theoretical Chemistry vol 3*, H. F. Schaefer (ed), Plenum, New York, 1977

T. H. Dunning Jr., *J. Chem. Phys.*, **55** (1971), 716

H. Eyring and M. Polanyi, *Z. Phys. Chem.*, **B12** (1931), 279

B. Finkelstein and G. Horowitz, *Z. Physik.*, **48** (1928), 118

V. A. Fock, *Z. Physik.*, **61** (1930), 126

M. J. Frisch, J. A. Pople and J. S. Binkley, *J. Chem. Phys.*, **80** (1984), 3265

S. Green, *Adv. Chem. Phys.*, **25** (1974), 179

D. R. Hartree, *Proc. Camb. Phil. Soc.*, **24** (1927), 89

D. R. Hartree, *The Calculation of Atomic Structures*, John Wiley, New York, 1957

W. Hehre, R. Ditchfield, L. Radom and J. A. Pople, *J. Am. Chem. Soc.*, **92** (1970), 4796

G. Herzberg and K. P. Huber, Molecular Spectra and Molecular Structure 4, in *Constants of Diatomic Molecules*, van Nostrand, Princeton, NJ, 1979

D. M. Hood, R. M. Pitzer and H. F. Schaefer, *J. Chem. Phys.*, **71** (1979), 705

H. M. James, *J. Chem. Phys.*, **3** (1935), 7

T. Koopmans *Physica*, **1** (1934), 104

W. M. Latimer and W. H. Rodebush, *J. Am. Chem. Soc.*, **42** (1920), 1419

B. Liu and K. Seigbahn, *J. Chem. Phys.*, **58** (1973), 1925

J. K. L. MacDonald, *Phys. Rev.*, **43** (1933), 830

A. D. McClean and G. S. Chandler, *J. Chem. Phys.*, **72** (1980), 5639

C. Moller and M. S. Plesset, *Phys. Rev.*, **46** (1934), 618

R. S. Mulliken, *J. Chem. Phys.*, **23** (1955), 1833

W. Pauli, *Z. Physik.*, **31** (1925), 765

L. Pauling, *The Nature of the Chemical Bond*, Cornell University Press, Ithaca, New York, 1939

R. Poirer, R. Kari and I. G. Czismadia, *Handbook of Gaussian Basis Sets*, Elsevier, Amsterdam, 1985

R. Renner, *Z. Physik.*, **92** (1934), 172

C. C. J. Roothaan, *Rev. Mod. Phys.*, **23** (1951), 69

B. Rosenblum, A. H. Nethercot and C. H. Townes, *Phys. Rev.*, **108** (1958), 400

S. Sato, *J. Chem. Phys.*, **23** (1955), 592, 2465

J. C. Slater, *Phys. Rev.*, **34** (1929), 1293

L. C. Snyder and H. Basch, *J. Am. Chem. Soc.*, **91** (1969), 2189

S. Weiss and G. Leroi, *J. Chem. Phys.*, **48** (1968), 962

D. N. J. White and M. J. Bovill, *J. Chem. Soc. Perkin 2*, (1977), 1610

H. Wind, *J. Chem. Phys.*, **42** (1965), 2371

Index